SAHEL WORKSHOP 1991

DEVELOPMENT EXPERIENCE IN THE SAHEL

PROCEEDINGS OF THE THIRD SAHEL WORKSHOP

7TH - 9TH OF JANUARY 1991

Introduction

This publication contains articles, presentation papers, abstracts and notes provided by the participants in the Third Danish Sahel Workshop held in Gilleleje in January 1991.

The programme of the workshop and a list of the participants can be found at the back of the publication.

So far, five Sahel workshops have been held in Denmark and a sixth workshop is due in 1994. The Sahel workshops gather researchers, development agencies, NGO's, private companies and others working with different diciplines in the Sahel or in similar arid and semi-arid areas of Africa and constitutes a forum where all interest groups can exchange experience and discuss the problems of the Sahel.

The overall theme of the workshop in 1991 was the experiences and strategies of the funding and executing agencies working in the Sahel: Is past experience taken sufficiently into consideration when preparing for the future?

No clear-cut answer was given to the question raised by the workshop. New development strategies do seem to take past experience into consideration. However, past experience is, although lived through, not always available! The "memory" of development agencies has often been too short. Data collected may not have been analyzed, or data may not even have been collected. The need for research in development assistance was the most apparent conclusion of the workshop.

The outline of the proceedings does not follow the programme of the workshop. The various contributions have been grouped in three main chapters according to their character. The contributions related directly to the overall theme of the workshop are found primarily in the first chapter covering issues of natural resources and socioeconomy, and in the third chapter which provides summaries of the presentations given by different development organizations. Chapter three also includes notes from discussions at the workshop. Chapter two gives abstracts and short communications of projects and activities of workshop participants.

The workshop was generously funded by Danida.

Lars Graudal and Holger Elme Nielsen

Humlebæk 1993

Contents

1. Presentations and articles

A multi-level approach to soil-erosion and land degradation surveys. Examples from the Lesotho lowlands

Lennart Strømquist

Abstract

Lesotho is a Southern African country which is characterized by severe land degradation caused by overcropping and concentration of the major agricultural activities to a rather small fraction of the country which is suitable for cropping, i.e. the Lesotho lowlands and Foothill regions.

A joint research programme initiated by the National University of Lesotho and Uppsala University, Sweden has focused on the development of methods for practical erosion inventories based on a geomorphological approach, including sediment budget analyses based on inventories, mapping and identifications of sediment sources, sediment routing and sediment storage areas. Processes which are well suited to monitor at different levels by inventories made at different scales. An increased understanding of the space and time variations of the geomorphological processes can hopefully give a better advice to the planning and conservation agencies optimizing their efforts in conserving the land in a most efficient way. The paper also refer to some related research made within current aid programmes on soil and water conservation.

Key words: Lesotho, erosion surveys, multi-level study approach, Southern Africa, South Africa.

1. Introduction

Regional background

Land degradation is a problem that has been observed Southern Africa since quite a long period of time. Already in 1908 Bradfeld noticed that the settlers "were scooping out the richest and most beautiful valleys, leaving them dry and barren" and that Southern Africa would become a land "ruined by constant trampling". A "Drought Investigation Commission" in South Africa identified the problem on a national scale in 1923 and the first soil conservation programmes commenced some 10 years later.

In South Africa the Botanical Survey (cf. Acocks, 1953, 1975) made a detailed field survey of the vegetation types in order to monitor trends in land degradation. The report is certainly one of the first reports on the "desertification" problem at a regional scale from this part of Africa. As probable causes

Acocks (1975 p 5) pointed out that much of the erosion damage in the country is the result of not allowing for the small variations in grazing management concentrating the animals to the areas of richer soil types, in valleys and depressions, which are characterized by more palatable vegetation. This "zonal selective grazing" in an aggravated form is evident throughout Southern Africa and is responsible "for the virtual disappearance of grass from the Karoo..." In a series of maps Acocks illustrate the spread of the "Karoo-bush" vegetation from a western location in the Cape province well into the Orange Free State with the limit of "Karoo patches" encompassing Lesotho in 1950.

In Lesotho a government commission, "the Pim-commission" made an inventory of the soil erosion status in 1935 (Beinhart 1984) and found that soil erosion was increasing due to over-population, overgrazing and overstocking. Recommendations from the commission led to intensive soil conservation programmes completed in the 1960's and later repeated by different aid organizations during the post-colonial times.

Botanical evidence of land degradation (desertification) as well as observations of increasing areas of eroded land have called for regional cooperation in managing the land degradation problem. The first organization, SARCCUS (Southern African Regional Commission for the Conservation and Utilization of the Soil) commenced its work after the second World war and later the SADCC (Southern African Development Coordination Council) countries (i.e. the Southern African countries excluding South Africa) have begun a similar cooperation programme on soil and water conservation

In November 1986 the SADCC Coordination Unit for Soil and Water Conservation and Land Utilization organized a seminar on monitoring systems for environmental control in Gaberone, Botswana (SADCC 1987).

The wide range of seminar inputs reflects both the regional need for environmental monitoring programmes, but also the difficulties in designing a regional programme and methodology for environmental monitoring. Questions like scale, accuracy, contents and applicability were problems discussed in relation to the design of a monitoring programme. A joint statement from the delegates, however, sorted out the most critical issues to be within the following fields; hydrology, vegetation and land use, soil erosion and land degradation and research in environmental monitoring.

Summarizing the results the conference agreed upon that the overall objectives for a monitoring system, which partly should be based on applied remote sensing, should be to:

- provide a synoptic view of the environmental conditions
- assess the present degree of degradation
- develop means of predicting future changes
- monitor long term changes and trends
- provide a quantified base for evaluation

The environment

The Kingdom of Lesotho is located between longitudes 27° OO'E and 29° 30'E, and latitudes 28° 30'S and 30° 40'S. The country is drained by three river basins, Caledon River basin in the west, the Makhaleng River basin
in the centre and the Orange (Senqu) River basin in the east and south. The country is normally divided into three physiographic units: The Mountain and Lowlands Provinces, separated by the Foothill Region.

The Lesotho lowlands form a narrow strip along the western and south-western parts of the country. The region is separated from the foothills by steep slopes and sandstone cliffs. The relative relief of the lowlands varies from 5 to 100 m. The major landforms are hills and dissected plateaux, separated by river valleys. The lowlands range in altitude from 1388 m at the outflow of the Orange River from Lesotho, to 1900 m above mean sea level. The climate of the lowlands is cool to hot with a mean annual temperature of 15° C. The winters however, often have temperatures below zero with frost during most nights. The mean annual rainfall of the lowlands ranges from 500 mm in the south to about 800 mm in the north. Most of this rain occurs during the months of October to April. The lowlands have a marked dry season centred around June and July.

The lowlands are characterized by Karroo sedimentary rocks, (shales, sandstones and mudstones) into which are intruded igneous rocks (dolerite). There is a pronounced relationship between rock type, main geomorphological units, soils, geomorphological process activity and possible land-use, a fact which is an efficient tool in air-photo interpretation and planning (cf. Schmitz 1980). The rocks are covered mostly by alluvium, colluvium and aeolian deposits.

The dominant soils of the lowlands are lithosols on the steep slopes and river terraces, fersiallic and ferrallitic soil on the plains and lowlands spurs, claypan soil and vertisols along the topographic depressions. These soils normally form toposequences that are related to the topography and geology of the region.

The indigenous vegetation of the lowlands is dominantly grasslands of *Themeda-Cymbopogon-Eragrostis*. The south facing slopes normally support groves of indigenous trees and shrubs. The valley bottoms, topographic depressions and river banks are often clad with trees, reeds and bushes. The indigenous tree vegetation of the country has been supplemented with woodlots and plantations around homesteads, villages and mission stations.

The lowlands form the major location of the settlements (villages and towns) in the country. They are extensively cultivated with pastures relegated to steep slopes, along the edges of the sandstone spurs and hillcrests. The areas between field boundaries and along the gullied areas are normally used for grazing.

The result of this heavy human and livestock population on the lowlands is extensive erosion and overgrazing, and rapid degradation of both water and soil recourses of the region. The result of this intensive erosion is the high rate of siltation of dams and ponds constructed to store water for both livestock and for domestic use.

Geographical background to the projects

In 1982, and independently of the SADCC conference, the National University of Lesotho and Uppsala University, Sweden initiated a joint research programme on environmental monitoring in Lesotho which has focused on the development of methods for practical erosion inventories, based on a geomorphological study approach. This approach includes sediment budget analyses, based on inventories, mapping and quantifications of the sediment sources, sediment rooting and sediment storage areas.

Processes which are all well suited to monitor at different levels by inventories made at different scales. An increased understanding of the space and time variations of the geomorphological processes can hopefully give a better advice to the planning and conservation agencies optimizing their efforts in conserving the land in a most efficient way.

In spite of its small size Lesotho (31 000 km²) is a nation with severe land degradation problems, which is most probably associated with historical changes of land use and concentration and increase of the population during the last century. About 2/3 of the total area is steeply mountainous a fact which furthermore concentrate the population to the lowland areas (below the 2000 m contour).

During the last century the the population doubled by 1950 to 660,000 making a density of 21 per km^2, and by 1970 it had increased to 920,000 with the density in the lowlands rising to 115 per km^2 (Rowland 1974). The present population is about 1.5 million.

The increased population has led to a continuous fragmentation of the land and increased grazing pressure, both factors that are accelerating the soil erosion risk and reducing the soil conservation potential. According to Nordstrøm (1988) about 34 % of the land surface was used for cultivation in 1988 (about 10300 km2) versus 8 % of the suface in 1878. As early as in 1954 it was stated that the saturation point was reached in the lowlands, and that cultivation was spreading into the former grazing areas up the steep slopes.

Since the mid-1930's the colonial and post colonial governments of Lesotho have attempted to implement soil conservation programmes at a national scale. The first programme was almost completed during the mid-60's, but before completion signs of neglect and erosion characterized the conservation areas (Chakela 1974). The lack of popular participation in the programmes and a lack of understanding of the geomorphological hazards are the most probable causes of the failure. A more detailed presentation of the history of soil conservation in Lesotho is presented by Chakela and Cantor (1987).

A major aim of the project is to increase the knowledge how to identify the most important sediment sources, sediment transport routes and sediment storage areas in relation to landforms and land units that are easily identified making the future soil conservation work more costeffcient by concentrating the initial work to the most erosive land units.

2. The multi-level study approach

In order to investigate and analyse the relations between geology, geomorphology, land use and land degradation, the project has collected information at different observation levels, however, linked to each other by common field control areas, reference data etc. (Table 1). Some of the listed projects are not a part of the joint research programme as such but have partly been carried out by the same scientists for applied programmes in soil and water conservation within the framework.

The studies on the *regional observation* level include erosion and erosion hazard mapping at a national scale, based on studies of satellite images and compilation of existing data of topography, rainfall, land use and soil erosion.

Table 1: Soil erosion studies at different levels in Lesotho

Observation level	Year	Study objective
Regional level:		
Strømquist et.al	1986	Satellite mapping of regional soil erosion intensity in the Lesotho Lowlands.
Strømquist et.al.	1988	Satellite mapping of regional soil erosion intensity (Spot based) in the Maseru area.
Chakela and Stocking	1988	Erosion hazard mapping of Lesotho.
Larsson and Strømqvist	1991	Handbook on a practical approach to satellite image analysis for environmental monitoring.
Local level		
Chakela	1974, 1980	Studies of sediment transport and reservoir sedimentation in selected catchment areas in the Lesotho lowlands.
Strømqvist et. al., Lundén et.al.	1985, 1986	Spatial and temporal variation in sheet and gully erosion.
Nordström	1986a 1986b 1988	Spatial and temporal variation on in gully development. Relations to land-use and population pressure.
Schmitz	1980, 1984, 1987	Land systems and correlation of landforms, geology and active processes.
Berding	1984	Practical application of Schmitz's studies for land-use planning.
Lundén et. al	1990	Testing tailored satellite imagery for erosion mapping by analyses of top-soil Cs content
Field level		
Rydgren	1985, 1986, 1989	Catchment and plot studies of soil loss in a soil conservation area
Kulander	1986	Wash erosion studies in relation to vegetation type and cover by wash traps.
Kulander	1986	Studies of sediment sources, routing and storage on fields in mini catchment areas.
Kulander and Strömquist	1989	Observations of sediment sources, rooting and storage on fields in mini catchment areas by studies of top-soil Cs content.

The *local observation* level includes catchment surveys based on air photo interpretation and field studies of sediment sources, erosion processes, land use and conservation potential. The catchment areas have been used as reference areas for studies made at the the regional observation level.

The *field observation* level includes observations of for instance sediment routing in mini-catchments or within plot size areas in order to analyse the importance of micro relief, vegetation cover land use etc.

2.1 Regional level

Chakela (1974, 1981) studied the regional variation of sediment yield in the Lesotho lowlands (cf. local level). His studies indicated a regional variation in erosion intensity between about 200 t/km2/year and 2500 t/km2/year but as the selection of study areas was dependant on the location of reservoirs and water gauging stations the study did explain the local causes of soil erosion but not the regional variation pattern.

Therefore, in order to identify the spatial distribution of the regional erosion pattern, Strömquist, Lundén and Chakela (1986) produced a provisional soil erosion map of the Lesotho lowlands in the scale of 1:250,000. The map was based on visual interpretation of multitemporal Landsat MSS false colour composites. Surface (sheet) erosion was mapped in three classes related to the Southern African classification system (SARCCUS), developed for air photo interpretation of soil erosion.

The accuracy of the Landsat mapping was checked by detailed studies made in a number of control catchments throughout the Lesotho lowlands. As Lesotho has a good coverage of aerial photographs from the 1950's until present most of the field control was made by air-photo interpretation using the same classification system.

The test showed that it is practically possible to make a rapid regional appraisal of the erosion pattern by visual interpretation of satellite data. The spatial distribution of intense soil erosion was analysed by comparing it with digitized maps of soils and land systems, geology, vegetation, land use and topography (Bawden and Carrol, 1968). It was not, however, possible to make an acceptable correlation between the erosion pattern and any combination of the physical factors.

Recent studies by Strömquist, Larsson and Byström (1988) have furthermore tested the new SPOT satellite imagery for soil erosion mapping at both the regional and local levels. Two of the control catchments, which were used in the

Landsat study, were also used as test areas in order to classify soil erosion intensity by the original Southern African classification system. The results indicate an extremely good correlation between air-photo inter-pretation and satellite observations in the classes representing intense soil erosion, but less accuracy in the classes indicating low erosion. The inter-pretation scale of 1:400,000 in combination with a mapping scale of 1:100,000, and stereo viewing of the images made the Spot based erosion maps more accurate and easy to compare with base maps on soils and geology.

As the mapping of the extent of sheet eroded areas by satellite remote sensing, also gives us a synoptic view of the vegetation cover and soil moisture conditions during the individual years, it will also reflect the total sum of erosion events during long periods of time. Therefore the approach is difficult to use for a continuous direct monitoring of erosion changes.

Applying a method developed for Zimbabwe, Chakela and Stocking (1988) have recently produced a soil erosion hazard map of Lesotho, based on rainfall data, topographical information, vegetation cover and soil erodibility, all weighted into erosion hazard units based on the SLEMSA soil loss equation.

An interesting finding is that the most of the intensively eroded areas, as identified from the satellite images, are the areas characterized by the lowest erosion hazard.

By comparing the two maps we can, therefore, conclude that the actual regional erosion pattern has to be related more to the cultural factors (land use, land tenure, land use history, population density etc.) rather then to the natural factors favouring a high soil erosion, as indicated by the hazard map.

2.2 Local level

The catchment studies by Chakela (1981) indicated a difference in process activity between the study catchments. In order to investigate the temporal and spatial variation of erosion within the catchments Strömquist et.al. (1985) made a pilot study in the Ha Khitsane catchment near Mohales Hoek in the Southern part of the lowlands. The erosion mapping indicated a similarity in the position of the intensively sheet eroded areas and a spatial variation of gully erosion over time. Lundén et. al. (1986) applied the same approach in the other study catchments. One of the major conclusions from the studies (by Strömquist et.al. and Lundén et.al.) is that they question previous statements such as "the period of maximum gully activity is past" (Chakela 1981 p. 146). The results rather indicate a temporal variation between gully and sheet erosion as being the most important sediment source.

Nordström (1986, 1988) has followed up the catchment studies on gully erosion, especially the relationship between intrinsic and extrinsic gully forming processes and local thresholds responsible for the regional variation in gully erosion. Nordström (1988 p. 124) concludes that "the results of this study and a comparison with other studies of gully erosion in Lesotho indicate that the erosion cycles are highly local and that the thresholds of initiation are crossed at different times in different areas in the same region". Also at the local level the erosion pattern seemed to be more related to human factors rather than the natural erosion hazards.

Methods for erosion mapping at the local level have been developed by SARCCUS for use with conventional aerial photographs (cf. SARCCUS 1981, Strömquist et.al. 1985). The SARCCUS system is based on a classing of the erosion into 5 classes related to features observable on aerial photographs and in the field. As repeated aerial photography is quite expensive for monitoring purposes Strömquist and Larsson (1988) tested SPOT satellite imagery for erosion mapping also at the local level. It was found that a combination of multi-spectral (XS) images and panchromatic (P) images in a stereo model gave almost the same accuracy in erosion mapping the local erosion pattern according to the SARCCUS system as a mapping of the same area from colour air-photos in the scale of 1:20,000. The reason was the combination of the image colour infor-mation of the XS registration and the higher resolution of the P-registration.

Digital image processing has so far never found a practical application in erosion surveys for applied soil conservation. For comparative purposes, however the method was tested in Lesotho (Lundén, Strömquist and Nordström 1990). "Tailored" images were produced in the form of infrared colour composites with the outlines of the eroded areas mapped according to the maximum likelihood classification and by creating new imagery by principal component analysis. The results were tested by conventional image inter-pretation and by using the ^{137}Cs of the topsoil as an indicator of areas of soil erosion or accumulation (cf. field level).

In a pilot study of the applicability of the ^{137}Cs method to Lesotho environments in the Roma Valley (Kulander and Strömquist 1989) it was found that a pattern of erosion and deposition sites could be distinguished by values of ^{137}Cs loading. The same approach was used in testing the remote sensing study. In this case, however the samples were collected on an on/off basis, i.e. on distinct (on the satellite imagery and aerial photographs) eroded areas vs. less eroded land. This method gives less accuracy in the quantification but was the only practical way of checking the remote sensing data in a spatial concept.

The areas classified on the tailored satellite imagery represent areas of moderate to very severe erosion. The experiment showed us that it is possible to use this approach for erosion surveys but it also to the conclusion that in order to apply the method a knowledge of the specific local conditions is crucial. For practical purposes it is therefore advised to map and classify erosion from stereo SPOT imagery using the topographical model as reference.

2.3 Field observation level

Detailed plot scale field observations have become a classical methods in assessing soil erosion in different agricultural environments. Studies by Kulander (1986) and Rydgren (1986, 1988, 1990) aim at an increased under-standing of not only the variation in erosion but also to explain the field spatial and temporal variation in sediment dynamics as well as defining the most appropriate observation level for field observations.

Rates of erosion under natural forest vegetation are in many climatic zones considered as indicating a geologic norm of erosion which can be used as a standard for assessing the impact of various land use practices and for judging the severity of accelerated erosion. The natural vegetation in Lesotho is grassveld (Acocks 1975). To maintain a good vegetation cover tree plantation is encouraged for soil conservation purposes. In order to study how the local vegetation influence sediment production Kulander (1986) studied run-off and erosion under different types of vegetation cover. The erosion rates (on Albaqualfs respective Argiudolls soils) of two areas with undisturbed grass cover were used as a reference.

Table 2: Description of field experiment areas (From Kulander 1986).

	Area 1	Area 2	Area 3	Area 4
Covers	Grass	Grass	Pine	Eucalyptus
Soils	Rama	Tsilsi	Tsiki	Tsiki
Canopy cover %	-	-	35	20
Ground cover %	95	50	95	10
Type of cover	Grass	Grass	Litter	Litter
Slope(°)	5.2	3.9	3.6	6.4

Table 3: Range of sediment concentration in run-off (From Kulander 1986).

	Area 1	Area 2	Area 3	Area 4
Minimum sed. conc (g/l)	0.65	0.06	0.00	5.00
Maximum sed. conc (g/l)	1.73	2.32	9.38	51.3

Small scale wash traps of Gerlach trough types were installed on 4 areas to collect the run-off and sediments. The traps were 0.5 m wide. The study areas consist of irregularly shaped drainage areas to each wash trap. While this study was not aimed at quantitative measurements of soil loss but a ranking of fluvial transport, no plot boundaries were constructed. The results of the experiments are, however, surprising and might question the conservation value in many tree planting projects, as for instance the *Eucalyptus sp.* hinder the development of a protective field layer vegetation.

Studying the erosion of farmlands in the Maphuseng area in Southern Lesotho, Rydgren (1988, 1990) found that rainstorms exceeding 3-5 mm/h caused surface run-off from grazing land versus a storm exceeding 15 mm/h needed to create surface run-off from cultivated fields. In a following study Rydgren (1990) studied run-off within small catchment areas and from run-off plots. The plot studies were aimed at quantifying soil loss from different agricultural management practices, one "modern" with hybrid maize seed and fertilizer and one "traditional" with local maize seed. 12 plots were installed, 4 on each of the 3 soil types investigated. The plot size is 4.5 by 13.5 metres. In general the plot scale gives significantly lower erosion rates than the estimates from the subcatchment scale. The principal difference between the two scales is the low rate of linearized run-off in the plots, which according to Rydgren (1990) seems to indicate that high run-off leading to gully erosion is the reason for the higher catchment figure at least during his observation period.

Kulander (1986) also observed the local sediment sources routing and deposition by relating the processes to the geomorphological units described by Schmitz (1980, 1984) in order to identity the role of the various land-units as sediment sources or sediment storage areas. Difficulties in direct field monitoring introduced the ^{137}Cs content of the top-soil as an indicator as the removal or redistribution of the Caesium adsorbed to the soil particles is related to the erosion cycle cf. Kulander and Strömquist, 1989).

Table 4 (see below) summarizes the observation made of the top soil Cs content in a small (0.5 km2) catchment area in Roma valley. It is interesting to note the erosion of the debris slopes, the net storage of sediments within the foot-slope, the lack of storage of fine grained material in the gullies and the deposition in the reservoir. The annual sediment yield to this resevoir was 1700 t/km^2 between 1973-1977, and 984 t/km^2 for the period 1977-1986. A possible application of the ^{137}Cs studies is practical advice to conservation and planning programmes concerning erodibility and conservation potential of certain land-units, hence reducing the total cost for catchment conservation.

3. Conclusions

The Lesotho study has clearly illustrated the need for a multi level study approach for a full understanding of the landscape, its conservation potential and sensitivity to recent environmental changes.

The erosion studies made at the regional scale clearly illustrate how a national erosion pattern is more related to historical and land use factors, rather than of the physical parameters responsible for the spatial distribution of high risk areas. The botanical observations of land degradation refered to at the beginning of this paper furthermore relates the Lesotho land degradation pattern to a far larger area and man induced desertification trends rather than observed climatic fluctuations or spatial variation of geographical or geological parameters.

Table 4. Relations between geomorphology, land-use and dominant geomorphological process and top soil Cs content observed 1987 in a Roma Valley mini-catchment

Land unit/land use	Loading (mB/cm2)	% of reference area	Process
Reference area/maize	30	100	"inactive"
Debris slope/grazing	20	66	Erosion
Footslope/grazing	142	473	Storage, deposition
Accumulation	84	280	Storage, deposition
Glacis/ cultivation			
Gullied land/intensiv grazing	0	0	Erosion
Reservoire	77	(257)	Deposition

The local (catchment) level has been most informative on the influence of the physical (i.e. geological, topographical, pedagogical etc.) factors and their influence on soil erosion and sediment transport as well as the relations between human impact and natural processes. The observed variations in gully development between the different study catchments clearly illustrate the influence of man

The field scale illustrates the importance of detailed field studies and quantifications to explain the impact of conservation practices, land-use and vegetation cover but also how they can be used to identify land-units which are important indicators of the processes in order to prevent further accelerated erosion and advice on possible land-uses and conservation strategies.

The combination of the three observation levels has furthermore increased the knowledge of the process response to human impact and natural processes to a far larger extent than had been possible by only using a single level approach.

References

Acocks, J.P.H..l953; Veld types of South Africa. Memoirs from the Botanical Survey of South Africa No 28.160 pp

Acocks, J.P.H..1975; Veld types of South Africa. Memoirs from the Botanical Survey of South Africa No 40.170 pp

Bawden, M.G and Carrol, D.M.,19G8; The Land resources of Lesotho. Land resource Study 3, Land resources Division, Department of Overseas Survey.87 pp.

Berding, F.R.1984; Suitability calssification of soils and climates for a number of land uses in Lesotho. Institute of Land-use planning. Min. of Agriculture and Marketing. Lesotho, 126.

Beinhart, W.,1984: Soil erosion, conservation and ideas about about development. A Southern Airican exploration,1900-1960. Journal of Southern African Studies, Vol 11:1.

Bradfield, E.R., 1908; Erosion and Dessication of the Karoo. The Agricultural Journal of the Cape of Good Hope., Vol 33:2.

Chakela, Q.K., 1974; Studies of soil erosion and reservoir sedimentation in Lesotho. UNGI report 34. pp 479-495.

Chakela, Q.K., 1981; Soil Erosion and reservoir Sedimentation in Lesotho. UNGI report 54,150 pp.

Chakela, Q.K. and Cantor, J., 1987; History of soil conservation and soil conservation policy in Lesotho, Southern African Development Coordination Conierence, Report No 8,29 pp, Maseru, Lesotho.

Chakela, Q.K., Lundén, B. and Strömquist, L (Eds). 1986; Sediment Sources, Sediment residence Time and Sediment TransferCase Studies of Soil Erosion in the Lesotho Lowlands. UNGI Rapport 64,167 pp.

Chakela, Q.K. and Stocking, M.1988; An improved methodology for erosion ha zardmapping. Part II: Application to Lesotho. Gegraiiska Annaler Vol 70A:3, pp181-190

Gerlach,T.,19G7; Hillslope troughs for measuring sediment movement. Revue de geomorphologie Dynamique,4:173

Kulander, L.,198G; Sediment Transpoffi under Different Types of Vegetation Lesotho, UNGI Rapport 64, pp 95-101.

Kulander, L. and Strömquist, L. 1989; Exploring the use of top soil 137Cs content as an indicator of sediment transfer in a small Lesotho catchment. Zeitschriit fiir Geomorphologie, Neue Folge, vol 33, pp 455-462.

Larsson, R.& and Strömquist, L.,1991; Handbook on a practical approach to
    satellite image analysis for environmental monitoring. SADCC coordination
    unit for soil and water consenation and land uilization, Maseru (in press)
Lundén, B., Strömquist, L. and Chakela Q.,1966; Soil Erosion in Different
    Lesotho Environments.-Rate and sediment sources. UNGI Rapport 64, pp 33-
    47
Lundén, B.,Strömquist, L. and Nordström, K.1990; An evaluation of soil
    erosion intensity mapping from SPOT satellite imagery by studies of colour air-
    photos and top soil content of [137]Cesium. UNGI Rapport 74, pp 13-37.
Nordström, K., 1986; Gully Erosion In relation to Extrinsic and intrinsic
    Variables. UNGI Rapport 64 pp.49-68.
Nordström, K., 1988; Gully Erosion In the Lesotho lowiands, A geomorphological
    study of the interactions between intrinsic and extrinsic variables. UNGI
    rapport 69, 144 pp.
Rowland, J.W. (Ed) 1974; The conservation ideal. Being the SARCCUS record
    for the period 1952-1970. Southern African Regional Commission for the
    Conservation and Utilisation of the Soil. Pretoria.378 pp.
Rydgren, B.,1986; Soil Erosion in the Maphutseng and Ha Tabo Soil
    Conservation Areas. UNGI Report 64, pp l03-l20.
Rydgren, B.,1988; A geomorphologicai approach to soil erosion studies in
    Lesotho. Geograiiska Annaler Vol 70A:3, pp 255-263
Rydgren, B.,1990; A geomorphological approach to soil erosion studies in
    Lesotho. - Case studies of soil erosion and land use in the southern Lesotho
    Lowlands. UNGI report 74, pp39-89.
SADCC 1987; Monitoring systems for environmental control. Report from as
    seminar held in Gaberone. November 3-7,1986. Southern African Development
    Coordination Council (SADCC). Coordination Unit for Soil and Water
    Conservation and Land Utilization, Maseru, Report 13,2452 pp.
SARCCUS 1981; A system for classification of Soil Erosion in the SARCCUS
    region. Compiled by the SARCCUS subcomittee for Land Use Planning and
    erosioncontrol. Department of agriculture and Fisheries, Pretoria, 20 pp.
Schmitz, G., 1980; A rural development project for erosion control in
    Lesotho, ITC Journal 2 pp 349-363.
Schmitz, G.(Ed), 1984; Lesotho, Environment and management. Geography
    Department, National University of Lesotho. 150 pp.
Schmitz, C. and Royani, F., 1987: Lesotho, Geology, Geomorphology, Soils.
    National University of Lesotho 1987,201 pp.

Strömquist,L. Larsson, R-&, and Byström, M. (Eds),1988; An Evaluation of The Spot Imagery Potential For Land Iiesources Inventories and Planning. A Lesotho Case Study. UNGI report,68,44 pp

Strömquist, L, Lundén, B. and Chakela Q.,1986; A Soil Erosion Map of the Lesotho Lowlands. -A Case Study using visual interpretation of multi-temporal Landsat False Colour Composites. UNGI Rapport,64, pp 15-32

Strömquist,L., Lundén,B. and Chakela, Q., 1985; Sediment sources, sediment transfer in a small Lesotho catchment.-A pilot study of spatial distribution of erosion features and their variation with time and climate. South African Geographical Journal, 67:1:3-13.

Desertification control on sandy soils in Sahelian West Africa with special reference to Mauritania

Axel Martin Jensen

Summary

Drought has caused a precipitation loss of 37 % in the desert, about 50 % in the former Saharo-Sahelian zone and 43 % in the true šahelian zone. In the ancient desert, human activities have been adapted to desert conditions for centuries and little degradation is observed. Drifting sand is a severe problem in the former Sahelian zone, which now has a desert climate. Barcanes move at high speed, whereas an eolian dynamic in the form of linear dunes form a more stable landscape. Sand movement can hardly be stopped, but the movement of barcanes and linear dunes can be checked, essentially by different pallisading techniques. The supply of material for the protection of the socio-economic infrastructure, may be a problem in this zone. In the actual Saharo-sahelian zone, irreversible degradation may amount to 20 % of the surface. Deep plantings of *Prosopis juliflora* and *Leptadenia pyrotechnica* without watering have succeeded in stabilizing drifting sands. If dunes are lower than 2 m, the survival rate may be higher than 70 % and ligneous production of Prosopis has been measured from 0.6 to 4.8 m3/ha/y depending on the precipitation and fertility of the sand. In the true Sahelian zone irreversible degradation is sporadic and rehabilitation may be obtained by sylvo-pastoral management only. In the Sudano-Ssahelian zone irreversible degradation is rare. Here the yield of dry farming may be increased by agro-forestry based on *Acacia albida*. Farm trees of *Acacia albida* managed in a high forest system are estimated to produce 1.8 m^3/ha/y.

In 9 of the most important regions in Mauritania, the cost of desertification control has been estimated to 10 US $ per inhabitant per year .

The last chapter deals with bottlenecks in local institutions and donor policies.

1. Ecological Problems and Technical Solutions

1.1 The Drought

Table 1 compares the precipitation during the drought 1971-87 with the precipitation of 1941-70. The loss has been most severe in the former Saharo-sahelian zone (P 100-200 mm) which has lost about 50 % in annual precipitation and consequently about 150,000 km² have shifted into a desert climate. The former desert (P < 100 mm) has been less affected, whereas the former true Sahelian zone (P 200-400 mm) has suffered a loss of 130 mm/y corresponding to 43 %.

Table 1: Precipitation (P) loss during the drought 1971-87 in Mauritania

CLIMATIC ZONE*) 1941-70 (Stations)	PRECIPITATION		LOSS
	1941-70 mm/y	1971-87 mm/y	%
DESERT, P < 100 mm (Bir-Moghrein, Zouérate, Chinguetti, Nouadhibou)	50.7	31.8	37
SAHARO-SAHELIAN, P 100-200 mm (Adrar, Nouakchott, Tidjikja, Boutilimit)	152.2	77.1	49
TRUE SAHELIAN, P 200-400 mm (Moudjeria, Tamchakett, Aleg, Kiffa, Aïoun el Atrouss, Timbedgha, Nema, Rosso, Boghe, Kaedi)	309.0	176.8	43

*) According to LE HOUEROU (1979)

As biomass production is closely related to precipitation, these seventeen years of drought have caused hardly reversible changes. In 1965, 73 % of the Mauritanian population were nomads, in 1988 only 12 %. Sanding up of social and economical infrastructures is a general phenomenon in Mauritania today.

This chapter will analyse desertification problems on sandy soils in West Africa and propose technical solutions concerning the following four climatic zones: the desert, the Saharo-Sahelian zone, the true Sahelian zone and the Sudano-Sahelian zone.

Figure 1 presents actual isohyets on a simplified soil map of Mauritania. In the following, all climatic figures will refer to the drought period.

Figure 1: Simplified soil map of Mauritania with isohyetes refering to the drought period

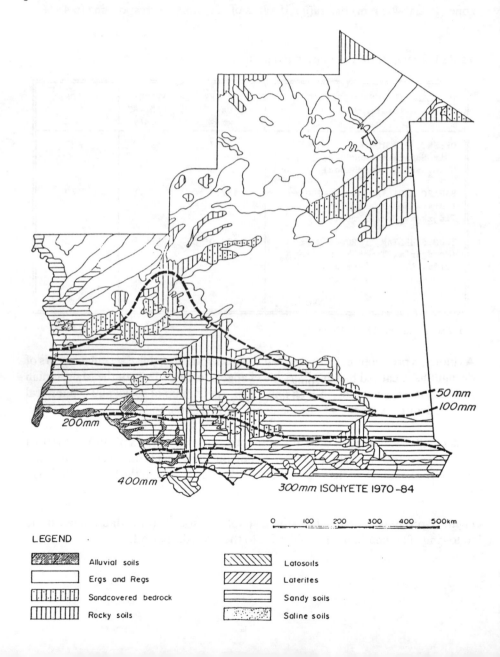

LEGEND

Alluvial soils	Latosoils
Ergs and Regs	Laterites
Sandcovered bedrock	Sandy soils
Rocky soils	Saline soils

1.2 The Ancient Desert Zone

Desertification in the desert is a pleonasm. As a matter of fact in the areas which have been desert for centuries, life adapted to the desert conditions long ago. Nomadic migration follows rainfall. Where pastures used to be available every third year, they may now be available only every fifth year, but the mode of exploitation has not changed. Oases are located along permanent water resources usually emerging along the contact between "ergs" (vast spaces of living sands) and mountains e.g. Ouadane and Tichit. Although close to shifting sands these oases have survived in the desert for centuries because the movement of the sands is controlled by topography. Generally the ecological balance has not been much disturbed in this zone, and beautiful forested areas of *Acacia tortilis* situated in sand covered depressions may be found e.g. in the vicinity of Guelb er Richad. These Saharan forests are often exploited by pollarding in a sustained manner either by the nomads or by professional charcoal burners from the oases.

1.3 The New desert zone

This zone, which used to receive more than 100 mm of rain per year but now receives less, is the zone which has suffered most from desertification because the combination of climatic changes and the actions of man has destroyed the vegetation.

Tree species such as *Acacia senegal* and *Commiphora africana* have disappeared from this zone like many annual grasses, while, due to sand drift the perennial desert species like *Calligonum comosum, Leptadenia pyrotechnica, Aristida pungens and Panicum turgidum* have not yet had the time or the possibility to invade these areas.

Barcanes at Nouadhibou have been measured to move at the speeds shown in table 2.

The movement of a 10 m high barcan south of Atar has been measured to 25 m per year. However on sandy substrate and with changing wind directions, as common in Mauritania, sand does not accumulate in barcanes, but in linear dunes.

The movements of a 4 m high linear dune has been measured to 15 m per year in southerly direction, absorbing most of the eolian energy in an East-West oscillation on the spot, with an 18 m amplitude of the shifting dune crest.

Table 2: Barcan movements at Nouadhibou (COURSIN 1954)

HEIGHT OF BARCAN m	SPEED m/y
3	88
4	70
10	38
16	22

The pattern of air movement around linear dunes does not permit the existence of sand accumulation close to it. Thus linear dunes form a rather stable landscape where vegetation may become established in between the dunes, whereas barcanes due to their speed are much more difficult to control.

Topography also has an influence on sand movement, usually by keeping living dunes confined to certain areas as for instance the echo-dunes on the windward side of inselbergs, which despite impressive shifting sand masses do not threaten oases situated on the leeward side of their base.

In a desert environment sand movement can hardly be stopped, but the movement of dunes may be checked.

KERR & NIGRA (1951) give examples on how to destroy barcanes or deviate sands threatening Standard Oil installations in Saudi Arabia. This experience may be extrapolated to environments subjected to a barcanic type of eolian dynamics.

In Mauritania, linear dunes have been checked by pallisading in a chessboard pattern or more economically by building up a fore-dune perpendicular to the linear dunes.

Figure 2 shows the effects of a 30 month old fore-dune protecting a road close to Nouakchott. The wind will be checked by the fore-dune and consequently drop the sand it carries. At a certain distance from the fore-dune the wind will speed up again and recharge with sand. If the fore-dune is erected in such a distance from the road that it is situated in the zone of recharging, the wind it self will clear the road of dunes.

Figure 2. Profile of a 30 month old foredune protecting a road close to Nouakchott.

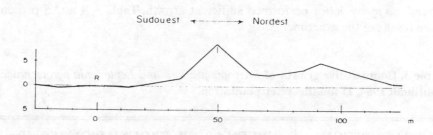

Planted trees and shrubs may survive under 100 mm of yearly precipitation, but they will never be able to take over the protection. Thus under less than 100 mm of rain, anti-eolian erosion measures have to be mechanical and must be maintained continually.

Under desert conditions it may be problematic to find materials for the mechanical defense systems. Branches of *Balanites aegyptiaca* or *Leptadenia pyrotechnica* have to be transported over long distances. If water resources are available materials may be cultivated. This is evident in the oases, where the palm leaves constitute an excellent material.

Oases in Mauritania have in many places been evaded by sand from the batha (wadi) along which they are situated. With the decline of run-off removing the sand in the batha, this type of sanding up has accelerated. Systems to prevent this and to clear the oases of the sand already there, have to be experimented with.

Where palm leaves are not available *Sesbania sp.* (e.g. sesban, pachycarpa) may be cultivated. They make an excellent fodder and in Mauritanian desert conditions (Akjoujt, Nouakchott) when irrigated on desert sand they become 2.5m high within 3 months. In Inchiri, the stems have shown resistance to sand abrasion for over a year.

1.4 The Saharo-Sahelian Zone

In the Saharo-Sahelian zone (P 100-200 mm) experiments carried out in Boutilimit (P 1971-87 99.3 mm) have shown that a biological stabilization of shifting sands is possible provided that living dunes are less than 2 m high. Clay content of dune

sand was 0.7 % and the interdunal soils, constituted of the eroded surface of the
fossil ogolian dunes, hold a clay content of 1.1 %. *Prosopis juliflora* showed good
growth on recent sand deposits and little growth on interdunal soils, where
Leptadenia pyrotechnica performed sufficient growth. Table 3, 4 and 5 present
some results of the experiments.

Table 3: Comparative growth of *Prosopis juliflora and Leptadenia pyrotechnica.*
Boutilimit 1984, 11 months after plantation.

SITE CONDITIONS	GREEN ABOVE GROUND BIOMASS kg/ha	
	Prosopis juliflora	Leptadenia pyrotechnica
Shifting dunes < 1.5 m high	1544	718
Denuded ogolian dune	165	539

Table 4: Planting techniques of *Prosopis juliflora.* **Boutilimit 1985, 13**
months after plantation.

PLANTING TECHNIQUES	SURVIVAL %	HEIGHT m
1.25 l containers with surface sand from underneath acacias. 3 month old and 50 cm high. Root collar placed in the soil surface. (Normal planting technique)	39	0.66
2.0 l containers with surface sand from underneath acacias. 4 month old and 90 cm high. Root collar placed in following depths:		
- 10 cm	87	0.83
- 20 cm	99	1.04
- 30 cm	86	1.06
- 40 cm	80	1.21

Table 5: Planting techniques of *Leptadenia pyrotechnica*. **Boutilimit 1985, 14 months after plantation.**

PLANTING TECHNIQUES	SURVIVAL %	GREEN ABOVE GROUND BIOMASS kg/ha
Containers with pure dune sand. 4 months old, 62 cm high. Root collar placed in 10 cm depth.	93	372
Containers with dune sand amended with complete slow liberating fertilizer. 4 months old, 83 cm high. Root collar placed in 20 cm depth.	94	1128

The essential learning from these experiments is that shifting sands may be revegetated without mechanical protection prior to afforestation using fast growing tree and shrub species. The technique consists of using big plants and placing the root collar deep in the moist sand. The top should preferably be 50 cm above ground to avoid abrasion. The advantage of this technique is that it will take time before the sand drift exposes the roots or covers the top, and that the roots are placed in a moist soil horizon so that the plants do not need watering. The season after the rains is often calm, so the fast growing species have time to establish themselves before the sand storms set in. Stands of *Prosopis juliflora* established in this manner have performed good growth:

- Boutilimit (P 1971-87 99,3 mm) shifting sands (clay 0.7 %) with dunes up to 2 m high and ground water table in 60 m depth. Age 31 month, survival 70%, height 2.4 m, total volume 1,5 m³/ha, above ground dry matter biomass 2.1 t/ha.

- Magta Lahjar (P 1971-87 150 mm) shifting sands (clay 3.1 %) with dunes up to 2 m high and ground water table in 8 m depth. Age 49 month, survival 83 %, height 4.8 m, total volume 20.2 m³/ha, above ground dry matter biomass 19.6 t/ha.

The sandy soils in the Saharo-Sahelian zone are vulnerable and often destroyed by a desertification process where the prominent features are drought - > decline in productivity - > over stocking - > destruction of vegetation - > wind erosion - > dune formation.

In 1979 in the Trarza region 18 % of the surface of the sandy soils was already covered by shifting sands (USAID 1982). Regeneration of this area can hardly be obtained without artificial revegetation, which in some areas may even result in an interesting forest production.

1.6 The True Sahelian Zone

In the true Sahelian zone (P 200 - 400) desertification in the form of living dunes is mostly concentrated around villages or in exposed areas. Elsewhere in this zone, the power of natural regeneration is so strong that it is generally sufficient, even in apparently very degraded areas, to protect only against grazing animals for a period of 2-4 years to restore the vegetation cover, at least of annuals.

The evolution of the afforestation system around the oasis Guérou (P 1978-87 180,8 mm) may be illustrative: Naked soil with some *Balanites aegytiaca* and traces of annuals was afforested and protected in 1984. Already with the rains in 1984 annuals cropped up and by the end of the dry season 1984/85 mortality was high among the planted trees. With the rains in 1985 the annual vegetation became abundant, and after three years, little of the original plantation had survived, but the soils were completely stabilized by the natural vegetation.

All over the true Sahelian zone many similar examples may be quoted. One of the most spectacular is the natural regeneration of the living dunes around the town Maïne Soroa close to Lake Tchad in Niger, which within 4 years became completely covered by grasses due to fencing only.

In areas with low pressure from grazing animals the tree vegetation suffers from the drought because the annual vegetation consumes the water before it reaches the root zone of the trees. On the other hand abundant natural regeneration of *Acacia senegal* may be observed in degraded areas following a decrease in animal pressure and a year with good rains.

A key word for sustainable exploitation in the true Sahelian zone is sylvo-pastoral management. Unfortunately very few convincing examples of engaging the local population in silvo-pastoral management are known. Although the benefits of protection are wellknown, voluntary communal protection without costly fencing has rarely been practised until now. An integrated approach where investments in animal health, wells, social services etc. are conditioned by the accept and the respect of temporary protection of certain degraded areas, may be a way out of status quo.

Around villages and towns, living dunes will often appear due to overgrazing and fuelwood collection. Afforestation of these areas is usually successful, but outside the dune areas afforestation will not succeed without careful weeding and even recharging of soil water by one or two years of fallow prior to plantation. The ligneous production of *Prosopis juliflora* may be estimated to 2 m^3/ha/y.

It is expected that villagers may collect fuelwood in the nearby surroundings without causing desertification, if the number of inhabitants is less than 500. On the other hand conversion to fossil fuel is conceivable in townships with more than 1500 inhabitants. This makes the villages with 500 to 1500 inhabitants priority targets for fuelwood plantations.

1.7 The Sudano-Sahelian Zone.

Only a small part of Mauritania is situated in the Sudano-Sahelian zone (P 400-600 mm) and sandy soils in this area are rare. The following is therefore based on experiences from Senegal.

With an annual rainfall above 400 mm a year, dry farming is possible. Desertification in the form of shifting sands is rare, but the zone suffers from loss in soil fertility due to less and less fallow and reduced areas around villages for the animal husbandry which produces the manure needed to restore soil fertility.

Desertification control is essentially based on agro-forestry. Agro-forestry is in many areas the traditional way of cultivation. The use and the silviculture of *Acacia albida* is wellknown (CTFT 1988, FELKER 1978, Giffard 1974, PELLISIER 1966, 1980) but other species such as *Butyrospermum parkii, Cordyla pinnata, Parkia biglobosa, Tamarindus indica* are also important although there have not been the same evidence of soil improvement as for *Acacia albida*.

It is characteristic that all these species are rare or non-existent in the natural forests in the area. They depend on man. The traditional silviculture consist of protection of young seedlings in the fields and pruning during the first years to achieve a straight bull.

Figure 3 presents growth of *Acacia albida* in Fatick, Senegal. Diameter and height has been measured and the age of the tree was known by the farmer who had cultivated the tree. This technique of measuring growth of farm trees has worked for many other species (*Borassus aethiopum, Cordyla pinnata, Diospyros mespilliformis, Ficus gnaphalocarpa, Sclerocarya birrea*).

Figure 3: Height and diameter growth of *Acacia albida* in Fatick, Senegal.

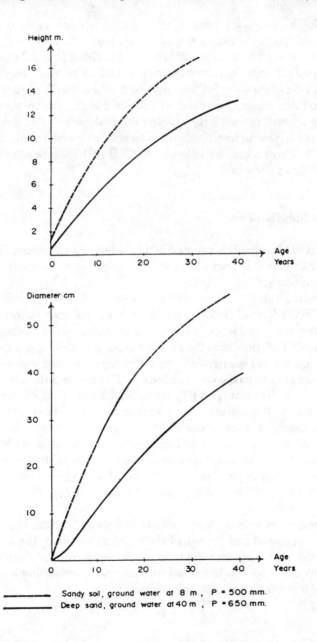

Sandy soil, ground water at 8 m , P = 500 mm.
Deep sand, ground water at 40 m , P = 650 mm.

Changing the management of the farm trees of *Acacia albida* from a selective cutting system to a high forest system may, according to the measurements, result in the following production:

- Districts of Diakhao and Niakhar. Rainfall 1971-81 app. 500 mm/y.

Deep sand (clay 2-3 %), ground water in a depth of 40 m.
High forest in a 40 year rotation.

* Production of branch wood (d < 10 cm)	20 m³/ha
* Production of stem wood	53 m³/ha
* Total production per year	1.8 m³/ha/yr

- District of Fimela. Rainfall 1971-81 app. 650 mm/y.

Sand with clay in 2 - 5 m depth, ground water in a depth of 8 m.
High forest on stump in a 35 year rotation.

* Production of branch wood (d < 10 cm)	42 m³/ha
* Production of stem wood	122 m³/ha
* Total production per year	4.7 m³/ha/yr

If farm trees of *Acacia albida* produce more than 1 m³/ha/y the ligneous production surpasses the requirements of selfsuffiency.

The planting techniques of farm trees of *Acacia albida* are however not yet well mastered. Survival rate of farm tree plantations in the Fatick area was only 25 %.

2. Costs

It may appear from what has been exposed above that the technical solutions to check desertification on sandy soils in the Sahelian zone does exist.

5 year programmes for desertification control in the 9 most important regions in Mauritania have produced the costs presented in table 6. During these 5 years, all socio-economic infrastructures will be protected against sand drift, fuelwood plantations assuring selfsuffiency will be established for all villages with 500 to 1500 inhabitants on sandy soils in areas with more than 150 mm rainfall, and pilot

projects for integrated rural development based on silvo-pastoral management will
be initiated in all the regions.

Table 6: Cost of desertification control in Mauritania.

REGION	MEAN PRECIPITA-TION	POPULA-TION DENSITY	COST US$/hab./y
	mm/y	No/km2	
Adrar	65	0.4	14.2
Tagant	85	0.5	9.1
Trarza + Nouakchott	100	8.5	8.0
Brakna	180	4.2	13.1
Assaba	210	4.5	6.2
Hodh el Gharbi	200	2.4	12.2
Hodh ech Chargui	200	1.0	5.5
Gorgol	300	12.5	13.1
Guidimaka	400	15.0	8.0
MEAN (weighted)	183	6.5	9.2

The national product in Mauritania is at present 420 US $/hab/y. With an annual
investment in desertification control amounting to 2 % of the national product
during 5 years, the most urgent desertification problems may be solved in what is
the most severely affected country in the Sahelian zone.

3. Bottlenecks

If it is technically possible to revers the desertification processes on sandy soils in
the Sahelian zone, and if more funds are available than actually utilised, it may be
useful to discuss some of the bottlenecks.

3.1 National and local institutions.

Table 7 presents a hypothesis of what could be the evolution of project assistance
to desertification control.

Table 7: The estimated evolution of project support for desertification control projects (no. of X's signifies relative importance).

PHASE	DURA-TION	INPUTS		
		FUNDS	ADMINIS-TRATION	TECHNI-CAL
I Conception, training	3	X	XXX	XXX
II Consolidation and extension	5	XX	XX	
III Budget subventions	10	XXX	X	
IV Self-reliance	-			

Projects often continue as true duplications of the pilot phase project. Local involvement is often consultative, with little responsibility in technical matters and virtually no financial responsibility.

A second phase, where emphasis is put on the building up of responsible national and local institutions to widen the physical targets of the project, could be a solution. But in many West African countries this seems, for the time being, doubtful, due to the actual structural adjustment policies, which have weakened the national administrations.

If the techniques established during the pilot phase have to be generalized it is however necessary to work through local administrations and to handover technical and financial responsibilities. How to do this under present conditions, and at the same time assure that the project assistance will be mainly to the benefit of rural populations, requires rethinking of donor policies.

3.2 Running costs

It is usually easy to obtain funds for investments, but nearly impossible to get subventions to cover running costs of projects after the initial establishment phase. All over the Sahelian zone it is possible to observe the remains of former development projects just beside new projects with nearly the same goals. The dogma of self-support has often been applied on a too short term in rural development projects in the Sahelian zone, where the marketing of rural products is not always evident and where climatic variations can destroy incomes in several consecutive years. However, increase in agricultural production may in many cases be desirable to improve nutrition and self-sufficiency.

In cases where reimbursement of loans is an additional part of the running costs this element may be the real reason for project failure. Desert control projects are in the short and medium term intended to check a loss not to provide monetary incomes. Thus financing desertification control through rural credit systems should be exceptional.

3.3 Local traditions

The traditional societies have often been accused of being the principal agent in desertification by their uncontrolled overuse of natural resources.

Careful study of local practices often reveals sophisticated systems of natural resource management and traditional legislation of land and tree tenure. The fact that these legislations are built on an oral tradition makes them only more useful in societies with a high illiteracy rate.

Project documents often make an apocalyptic picture of actual conditions. More emphasis should be given to the determination of how the traditional practices may be amended by the techniques the project aims to introduce.

Bibliography

CTFT, 1988: *Faidherbia albida*. Monographie. Centre Technique Forestier Tropical, 94736 Nogent-sur Marne, France.

COURSIN, A., 1956: Etudes des Barkhanes à l'Est de Port-Etienne. Territoire de Mauritanie. Bureau Central d'Etude pour les Equipements d'Outre-Mer, Paris.

FELKER, P., 1978: State of the art: *Acacia albida* as a complementary permanent intercrop with annual crops. University og California, USA.

GIFFARD, P.L., 1974: L'arbre dans le Paysage sénégalais.
Sylviculture en zone tropicale séche. CTFT, Dakar.

LE HOUEROU, H., 1979: Le role des arbres et arbustes dans les paturages sahéliens. Dans le role des arbres du Sahel.IDRC-158f, Ottawa.

KERR, R.C. & NIGRA, J.O., 1951: Analysis of Eolian Sand Control. Arabian American Oil Company, New York.

PELLISIER, P., 1966: Les paysans du Sénégal. Imp. Fabrègue, St. Yrieix, France.
PELLISIER, P., 1980: L'arbre dans les paysages agraires de l'Afrique Noire.
En L'arbre en Afrique tropicale - la fonction et le signe. Cahiers ORSTOM, Série Sciences Humaines, vol. XVII, no. 3-4. Paris.
USAID, 1982: Inventaire des ressources du Sud-Ouest Mauritanien. Remote Sensing Institute, South Dakota State University, USA.

Village associations and the state in Senegal

Henrik Nielsen

During the last decade there has been a remarkable increase in the researchers' interest in peasant associations in Senegal. WB-analysts and hardcore marxists, western NGO's and the Senegalese state alike have for once unanimously welcomed these associations as one of the few new things in rural Sahel that calls for some optimism. People talk about "a silent revolution", or say that the peasants have taken their destiny in their own hands" and it has become popular to end evaluation reports with African sayings heard in the field like "if they stick together the ants can concur the elephant", etc.

It is not my intention to try to kill this fire optimism burning and everywhere pour more fuel on the current afro-pessimistic fire burning everywhere except on the national Senegalese newspaper Le Soleil. However, I know a few other African sayings that in my point of view illustrate the political powerstruggles in rural Senegal more clearly: "When the elephants fight the grass is trodden down", highlighting who the victimms of these struggles are, or "The goat grazes where it has been tied" referring to the situation of the state bureaucrate in a Sengalese context. Furthermore I think that when so different organisations and individuals serving very different interest groups suddenly seem to agree on an issue, then it calls for a certain scepticism.

I shall therefore attempt to raise a few questions concerning these associations, in the hope that this might help us to avoid seeing the peasants associations as yet another panacea to tutal development with the inevitable disillusionment following it. At the same time I think it is very to try to focus on the limits of the activities of the peasant associations in order to come up with some more realistic expectations of the peasants associations in the future, a field that is sometimes ignored.

Much of the work on Senegalese peasant associations is theoretically weak and ideologically biased. I find that it is still a very common error implicitly to assume that rural Sahelian societies should be egalitarian and homogenous in their socioeconomic set-up. There is as well a tendency to ignore the fact that antagonistic interests exist on micro as well as on macro level in what we call the rural development process.

As an example I would like to mention as Belloncle who enthusiasticly sees the peasant associations as a means for the peasants to resist the exploitation of the state through organising. As an old revolutionary I naturally have a certain sympathy for ideas of this kind, but I must admit on the other hand, that I find the

grounds on which his conclusions are based mistaken. According to Belloncle
Sahelian villages are especially well geared for peasant associations for the
following reasons:

- the fact that the peasants live together in villages facilitates cooperation within
  the peasant society including exchange of goods, services and information.

- Sahelian villages with their palaver-democracy have a very long tradition for
  democratic rule and decision-making.

- there is a widespread economic homogeneity in the sahelian villages, even
  access to land etc. because of very limited capitalist penetration.

In my opinion all these three points are wrong. The fact that the peasants live
together does not indicate that they work together. Belloncle argues that the
Ujamaa-system in Tanzania failed because the peasants there did not live in
villages, but were scattered around, and therefore they were not geared for
cooperation. You cannot argue that because something has not taken place where
a given factor is not present, it is bound to happen where the factor is present.
Northern european peasants in the 19th century did not live in villages, and it is
their model we still try to sell to the African peasants as the classic cooperation
model.

I also question the degree of democracy that exists in Sahelian villages, and
emphasize that this varies enormously from village to village within Senegal. Caste
systems, slavery, gerontocracy, tribute-systems, very limited room of action for
women etc. seem to be a vital part of the rural Senegalese legacy. Furthermore, the
fact that everybody may have the right to speak under the palaver tree does not
imply that everybody is listened to. During a RRA mission I joined in eastern
Senegal it was our intention to gather the villagers on the first day in a general
meeting to discuss the reason for our visit. A Senegalese colleague, however,
advised us not to do this, because the chief would then express the opinion of the
village, and thereafter the villagers would be extremely reluctant to express views
that did not correspond with his during the rest of our stay.

Concerning the third point about the economic homogeneity, it is wrong to say
that mere capitalist penetration has caused the social stratification, although it
often reinforces it. There has always been exploitation of labour in some form, this
is not exclusively a capitalist deed. Regarding the free access to land of the
peasants, this is also highly questionable. It existed when land was plentiful, and

the scarcities were in terms of labour. Now when these things change, this point is no longer valid.

Finally the thought of the peasantry and the state as two more or less monolithic blocks opposing each other is simplistic, a point I shall come back to later. But at least Belloncle insists that there exist different interest groups in rural Senegal.

It sometimes seems as if this were not the case, especially among the adherents of the popular participation talk. An example of such an approach is the "Segou papers" that came out after a tripartite meeting between the Sahelian state agents, donors and representatives of the leading peasant associations in Sahel arranged and later published by CILLS and the Club du Sahel. The paper describes how the participants "left rhetoric aside to speak in plain terms about where they had succeeded and where they had failed in their attempts to halt the relentless advance of the desert". The discussions were "landmarks in their efforts to achieve a common objective", but at the same time there was "no question here of restricting the room for manoeuver of any of the partners, nor of dictating what each of the group or players should or should not do." The donors, the state agents and the peasant associations are referred to as the three "families" determined to work with a concerted approach.

When you look at the Africa debate over the last 20 years, these conclusions are a bit surprising. In the 70's foreign aid was the spearhead of western imperialism. In the 80's the talk was about the hypertrophic and predatory African state. Now all of a sudden it seems that all antagonisms have vanished, and what is necessary is just that people sit down and talk things over. I am aware that the Segou-newsletters were not intended as scientific reports, and should therefore not be treated as such, but to me it seems very doubtful that "rhetoric was left aside" as claimed, considering this very superficial treatment of a very complex topic with considerable impact on power structures in Sahel.

I would find it interesting to clarify in what way there historically has been cooperation among peasants in a Sahelian context, and in what way it differs from the cooperation as it is known in a European context, which was also the model that was more or less imposed on the peasants with the Senegalese cooperatives after independence.

I think Dominique Gentil is right when he points out that in the traditional European cooperative model peasants have worked together with a common goal and democratically decided how to invest the surplus of their concerted effort. In Sahel however it is different. Cooperation here is more based on reciprocity where the concerted work is to the benefit of one single man. For instance a number of farmers decide that one day everybody goes to work on A's field. Next time they go to B's field a.s.o. This reciprocity is however rarely symmetric. The peasants involved in this sort of work exchange often find themselves working much more on A's field than on B's. The famous Senegalese groundnut marabout production system is to a wide extent based on this sort of asymmetric reciprocity. The classic studies by Cruise O'Brien explain how this system has become so widespread. The massive voluntary work of the peasants on the marabout's field to his benefit does not reflect any "false consciousness", but is rather to be seen as a way for the peasant to insure himself against disasters like drought, where the rich marabout is obliged to help the disciples that have worked with him for years.

It is in this context that we have to understand the upsurge of the senegalese peasant associations, not as emerging from an egalitarian and homogenous society, where the peasantry is in a process of class formation, but rather as emerging from a society based on asymmetric reciprocity and clientilism.

With the falling prices of groundnuts it has become increasingly difficult for the Senegalese state to extract a surplus from the agricultural sector. With enthusiastic encouragement from IMF and the WB this has led to a drastic disengagement of the state in the agricultural sector, leaving the peasants with very limited possibilities of achieving credits. The natural answer to this is creating peasants associations in order to attract outside support. And the success of the association is therefore measured by its ability to attract outside support.

It is therefore my unfortunately empirically unverified thesis that the peasant associations, desperate as their situation is will tend to organize along two lines:

1. The peasants organize not according to their normal social structure demanding according to what their needs are. They rather tend to create associations the way the donors would like to see them, expressing their needs in terms of what is believed possible to obtain, i.e. what is currently in vogue within the donor community - planting trees, womens organisations, improved stoves, family planning etc.

This leads to projects that will not be sustainable. Furthermore in the process we see the emergence of a new rural elite. The most succesful peasant associations are led by repatriated well-educated people who have returned to their villages with great knowledge of the donor community. Along with the massive donor support the peasant associations have become a means for upward mobility in the bureaucracy for political careers of these people who are rapidly mounting in the ranks of the ruling Socialist Party. With the growing influence of the peasant associations I therefore do not think as it has been proposed, that it is likely that these will be seen as an attack on the political hegemony of the state. More likely is a scenario supposing an amalgamation of these new leaders of the peasant associations within the existing hegemonic power structures.

2. On the other hand we might see a parallel process identical to the one observed in the state-cooperative period in the 60's and the 70's, where peasants were organized on the basis of traditional clientilist lines reinforcing traditional power structures. These cooperatives were quickly dominated by traditional leaders as chiefs and marabouts, and there is a natural explanation for this with our asymmetric reciprocity in mind.

When asking why it was always the rich and powerful that were placed as cooperative leaders you often get the explanation that this is because a rich man is less likely to "eat" the money of the cooperative, an almost irresistable temptation for a poor farmer in seasonally severe survival difficulties.

This might be true, but there are also other explanations. By supporting an important marabout in the election of a cooperative leader, the peasant had realistic hopes that this person would be influential having affectionate ties within the state bureaucracy, and thereby facilitate a trickling down of state funds to the local level.

At the same time, being cooperative leader could be very profitable. By receiving gifts or, if you wish, being corrupt favouring some farmers at the expense of others, cooperative leaders could make substantial amounts of money. Therefore it was not free to get a position as leader. You had to pay a high price to the farmers for being elected which only the rich could afford. Another example of reciprocity in the hierarchic Senegalese society. There is no reason to believe that these mechanisms will be suspended in the new peasant association set up.

Finally I would like to repeat that in my view, peasant associations still call for some optimism. Rural Senegal is not passibe and apathetic, and a development that turns out different from what the donors expected might not be so bad. What I want want to criticize is the widespread naivity with which the subject is often treated, which might lead to a repetition of the errors made earlier with the Sahelian cooperative movement followed by inevitable disillusionment and increased afro-pessimism.

Pastoral tenure systems in Senegal - a discussion of pastoral management and exclusion rights.

Kristine Juul

Introduction

The concept of "the tragedy of the commons" was first forwarded by Garett Hardin in 1968 as a discussion of the worlds population problem. A small paragraph devoted to the management of free pasture lands had very important impact on theories of management of African Rangelands. In this paragraph Hardin claimed, that if free access existed to grazing lands, it was to be expected that all herdsmen would try to keep as many cattle as possible without regard to the possible overgrazing problem. In other words a potential conflict existed between group interests and individual interests whenever land was common property and the herds were private. Hardins conclusion was therefore that "freedom in a commons brings ruin to all"[1].

The hypotheses forwarded by Hardin have been critisized by a large number of scholars stating primarily that what Hardin describes as common property in reality is an open access regime, characterized by absence of property rights. As in most rural societies some kind of management of the individual use rights to common resources prevail, the open access regime is supposed to be less common than assumed by Hardin. Nevertheless the majority of the opponents[2] still consider Hardins example as a valid critique of an open access regime.

This approach, as well as several more or less inherent concepts such as overgrazing, overstocking, range degradation etc., have continued to serve as justification for projects in agro-pastoral zones. The underlying idea is still the one

---

[1] Garrett Hardin, "The Tragedy of the Commons", Science, p. 1244, vol. 162, 1968

[2] Among others Cyriacy-Wanthrup and Bishop, 1975. "Common Property as a concept in Natural Resources Policy", Natural Resource Journal no.ls, and Bromley and Cernea, 1989: "The Management of Common Property Natural Resources, Some Conceptual and Operational Fallacies" 57, World Bank Discussion Papers.

expressed by the "tragedy of the commons": that degradation of rangelands is caused by people, not by natural and climatic factors, and that correction of the situation requires institutional changes i.e. land reform directed towards a higher degree of privatization. Much of the literature on pastoral management continues to point out this necessity to "replace randomness by orders (Bromley) and introduce/reinforce management systems build on rights of exclusion (Mortimore[3]).

The aim of this paper is to discuss some of the conseguenses of the introduction of exclusion rights (and some of their underlying assumptions?). Examples will be presented from Senegal where different types of livestock projects have tried to introduce new tenure systems. A comparison will be made to the traditional systems and to the actual tenure systems prevailing in most parts of the agro-pastoral zone of northern Senegal (the Ferlo).

Some examples of pastoral tenure systems in Senegal

Tenure is most commonly typologized as one of 4 possible systems:

1) Common property
2) Private property,
3) State property or
4) Open access.

Whereas common and private property both are based on some sort of exclusive rights, the matter is more complicated for the 2 other systems. Theoretically, rights of exclusion are implied in state ownership, but apart from cases such as national parks etc. state property per se seldom occurs, but is rather part of a mixture of different types of tenure implying limited possibilities of exclusion. This is the case in most pastoral zones where the state ownership to all non-registered land was introduced in 1964, assigning only use rights to individual citizens.

---

[3] Mortimore M, 1989: "Adapting to drought: Farmers, Famine and Desertification in west Africa", Cambridge University Press, Cambridge, p.227 "Mortimore nevertheless recognizes the difficulties related to the recognition of group rights, as it serves to undermine the basis of transhumance.

1. Common property

According to Bromley[4], common property represents private property for a group since all others are excluded from use and decision making and because the individuals forming the group are submitted to certain rights and duties vis-à-vis the group. It differs from the private property regime in that the user is unable to alienate or transfer either the ownership or the use of that land to another individual.

1.1 The traditional Fulani pastoral system

The traditional Fulani pastoral system existing in the semi-arid Ferlo region of northern Senegal before the establishment of the deep-wells in the late 1950s may be caracterized as some form of common property regime.

Because of the low level of the ground water (in most of the Ferlo between 200 and 400m), the region was only accesible in a few months during the rainy season where the cattle could exploit the natural ponds. Seasonal transhumance took place mainly beween the Senegal River Valley (the Waalo) and the interior country (the Dieri). In order to protect the animals from attacking game, groups of herders joined their herds in order to follow a fixed transhumance route and regaining the same rainy season camps every year.

The exploitation of grazing lands was loosely structured. Around the fields, restrictions were established (called houroum nguese -prohibition of the field) limiting the acces of livestock to the cultivated areas during the rainy season. Similar restrictions were set up for the grazing lands (hourum douroungal) reserving the access of pastures to the livestock belonging to the camp. Contrary to the houroum nguese, no fixed limits existed for the latter, the primary aim being to secure that the animals could graze an entire day without approaching the fields of a neighboring camp[5].

It is obvious that these restrictions were more directed towards a protection of the agricultural production (and avoiding fencing of the fields), than towards enforcement of priority rights to pasturelands or ponds. Nonetheless the

---

[4] Brooley and Cernea 1989 p.15.

[5] Barral, 1982: "Le Ferlo des Forages - Gestion anciennne et actuelle de l'espace pastorale" p.30-31, ORSTOM, Dakar

enforcement of houroums assigned certain exclusive rights to the regular users by limiting the settlements of strangers within the controlled areas. The system can therefore be characterized as a form of common property regime.

With the introduction of new digging technology in the late 1950's and the discovery of an immense deep water table (the Maestricien), it became possible to exploit the pastures of the interior country year round. As the new wells were dug and managed by the colonial government, the Fulani fractions could enhance no control over the use neither of the wells nor over the surrounding grasslands.

Notwithstanding Barral contributes the loss of the practises of houroum during the early 1960s more to an increasing nucleation or individualisation of the pastoral households than to the loss of (a fairly fragile) control over water resources. The disappearing of predators resulting from of the colonial strychnine campaigns in the 1950s, as well as the increasing competition in the access to pastures, led many herders to abandon the security of the larger camps to establish smaller camps spred out on a large area.

The increasingly anarchic settlement strategies as well as the slackening of group solidarity ties obviously led to a decline in the authority of the traditional chiefs. The vanishing away of the institutions traditionally in charge of surveying the houroums was further enforced with the 1964 land law introducing state ownership to all non-allocated land. This left the Pulani fractions with little or no authority to enforce exclusive rights.

Today, rights of exclusion are enforced in very few areas and only during certain seasons (to protect areas of gum arabigue production from marauding goats or as a regulation of the location of residential sites to asure equal acces for all to the dry season pastures).

1.2 The Bakel and Senegal Oriental Livestock Projects

Other examples of common property oriented regimes are the rangeland projects of Bakel (USAID) and the PDESO[6]. According to Bromley, the project was justified by "rapid resource degradation due to overgrazing, lack of protection and

---

[6]The "Projet de Dévelopement de l'Elevage du Sénégal Oriental" was designed in the beginning of the 1990s by The World Bank. The Bakel project was designed over the same model. Both were implemented in the late 1970s.

the disregard of traditional property rights of local pastoralists by incoming herders from the more arid northern Ferlo[7].

1.4 million ha was divided among groups of pastoralists constituted by the projects social staff[8]. In order to secure the rights of the pastoral units The World Bank persuaded the government to pass legislation granting each pastoral unit longterm rights over segments averaging 20.000 ha. each. The government also promised to use its authority when needed, to support these associations against non-members, e.g. intruders from the north, thus creating a mechanism to support the groups exclusion rights. As a compensation to the herding groups traditionally passing through the area on seasonal transhumance delineated corridors were established.

Later, it turned out that most of the villages were built along established transhumance routes. In the sense that (transhumant) herds preceded the (sedentary) peoples, the herders from the north still have some residual right in the area. It was therefore difficult for them to see why old migration routes and watering holes were being assigned to relative newcomers.[9]

Furthermore, the population in the Bakel project turned out to be Toucouleur agropastoralists to whom cattle keeping was only a secondary activity. Less labour was thus invested in herding the animals, the cattle being kept in a more hazardeous manner. Had this been recognized from the outset, it might have been possible to turn towards a more adequate project design oriented towards for example mixed farming.[10]

On the whole, the PDESO project is considered to be a relatively succesfull livestock project. The pastoral units have managed to coordinate the grazing and watering activities by administering grazing rotations and the ecological recovery of the rangelands has proved satisfactory[11]. Nevertheless, the litterature available does not clearly express what this recovery implies and to which extent it is to be attributed to the better rainfalls at the end of the project period.

---

[7] Bromley 1989 p.34.

[8] In the Bakel project villages constituted the pastoral unit.

[9] Bennet et al, 1986: "Land Tenure and Livestock Development".

[10] ibid.

[11] Bromley 1989 p.36

The evaluation report for the Bakel project[12] draws the attention to the fact that the amelioration of watering facilities through the construction of artificial ponds has in fact contributed little to increase the herd productivity. The group favoured by the project is primarily sedentarised and the introduction of the artificial ponds, making large pastures accessible to the cattle was not fully appreciated by the villagers. It turned out that the ponds could retain water only for a short period" The gain of accesibility of pastures was thus reduced" Furthermore they were located too far away from the villages to be properly used and maintained by the villagers. Finally the traditionnal organization of the villagers demonstrated a certain reluctance to participating in the communal activities concerning the management of natural resources[13].

2. Private property regimes

According to Bromley "private property is the legally and socially sanctioned ability to exclude others - it allows the fortunate owner to force others to go elsewhere. ... The difference between private property and common property is not to be found in the nature of the rights and duties as much as it is the number to which inclusion or exclusion applies."

2.1 Projet de Reboisement de la Zone Nord/Projet Sénégalo-Allemand

The "Projet Senegalo-Allemand" in Northern Senegal is an almost classic example of an attempt to privatize rangelands insofar as it promotes individual "leaseholds" with a sanctioned right to exclude others.

The project is located in Widou Tingoly in Western Ferlo in a zone where the agricultural production is limited almost exclusively to subsistance production. The years of poor rainfall have even pushed some pastoralists to give up their agricultural production entirely.

The project was originally designed as an afforestation project, but after the first years the project managers realised that a more integrated approach was

---

[12] A. Cheikh al. 1985: "Projet Elevage, Departement de Bakel: Rapport Final d' Evaluation". CRET, University of Michigan.

[13] ibid p.4ll.

needed. Consequently they changed from a mere enclosure/plantation strategy to a strategy which takes into account the needs of users of the ameliorated grazing lands. However, no in-depth study was made of the herding strategies and the production systems of the zone and during the first 10 years of the project no socio-economical personel was employed by the project.

The aim of the project is now to persuade the herders that having a limited but healthy herd with access to an adequate (if not abundant) amount of grazing lands is better economy than keeping a large herd. In order to demonstrate a more efficient model of range management, a huge area (approximately 20.000 ha.) surounding the deep-well has been divided into 6 "slices of cake". Each slice is subsequently to be subdivided into lots, limited by fences and "leased" to individual herders or groups of two. The state continues to be the formal owner of the land. In order to control the carrying capacity, each lot may only contain 40 cows, 40 sheep and 20 goats. As the herds grow in size the surplus is to be taken away from the lot.

It is undeniable that within the parcels, this procedure has had a very positive impact on the herds as well as on the vegetation. But outside this area, the project actually contributes to an increasing grazing pressure as the destocking from the lots, instead of being sold as originally assumed by the project personel, constitutes a rapidly growing herd grazing outside of the project fences.

Because of this "dual" herd structure, plans for the further extension of the project have encountered fierce hostility among herders already disposing of an allotment. For these people enclosure of the remaining "slices" around the central well will make access to water still more difficult for the "second" herd. Further limitation of the "open" grazing lands will also intensify labour problems as the family members in charge of the second herd will be forced to settle farther away from the allotment.

The reduction of grazing land as well as the difficult accessibility to watering facilities create even more problems when it comes to the transhumant herders or the agropastoralists owning only a small number of animals. For the transhumant herders the extension of the enclosed areas and the increased distances between the pastures and the watering facilities make Widou Thiengoly a less desirable halt on the traditional transhumance routes. For the indigenous population the adoption of stocking rates in the allotments have implied the exclusion of herd owners disposing of less than the prescribed number of cattle[14].

---

[14] This will be corrected in the prospected allotments where several smaller herders are to share one large enclosure.

Furthermore distribution of lots have been made with the assistance of leading members of the community. In fact family ties to members of the distribution committee have played an important role in the distribution of plots, and the project thus contributes to further social differentiation of the community.

The establishment exclusion rights goes hand in hand with the idea of sedentariness. When settled in his fenced allotment the "project-herder" no longer contributes to the herding solidarity necessary to maintain a more mobile system.

The fragility of this system became apparent as bushfires in November 1988 devastated some of the lots forcing the herders to move into neighbouring areas. Similarly a breakdown of the deepwell in April 90 led to the temporary resettlement of large number of herders in the vicinities of the neighbouring wells. As free access is no longer attributed to strangers wishing to settle in the project area, it is foreseeable that such intrusions in the neighbouring communities might create conflicts in the future. It may seem as if the design is only viable as long as the project personnel is present to patriarchally provide solutions for the families in the scheme. The herders seem to recognize this, as they continue to build up large herds as an insurance against future calamity instead of selling off their cattle as expected by the project staff. This inevitably leads to a lack of respect of the stocking rates enforced by the project.

2.2 Allocation of individual titles to herders

An interesting example of "spontaneous" privatization of grazing lands is the allocation of land to Fulani herders through the rural councils (Conseils Rurales). As the prevailing jurisdiction in Senegal fails to recognize grazing as productive land use (mis en valeur), land can in fact mainly be attributed to agricultural production.

Nonetheless singular cases of land attributions to pastoral production units have been observed in agro-pastoral areas by the north-eastern frontier of the peanut production zone[15]. In 'these areas, the expansion of mouride "dara's"[16]

---

[15] For example Thiel et Gassel in the department of Linguère. (The information stems from an interview with the Chef Secteur de l'Agriculture of Linguère, Nov. 1988.)

in search of new lands for peanut production contributes to a increasing presure on the land.

During 1988 a number of individual Fulani herdowners submitted requests, each of several hundred hectars of land, to the rural councils. Even though it was well known that the herders were unable to put these large areas under cultivation, the attributions were granted. The decisions of the Rural Councils were approved by the authorities (the Service Departementale de l'Agriculture of Linguère) who recommended that no attributions should exceed 20 ha. in the future, stating that the pastoral production units involved sought only to gain private ownership to communal grazing lands.

Nevertheless the positive reaction of the rura I councils, composed mainly by agropastoral Fulanis, should be interpreted rather as an act to halt the mouride expansion, than an attempt to privatize rangelands. It will be interesting to see, if these private attribution will have any implication for the future management of the lands, and to what extend these (illegal) attributions will be copied in the neighbouring communities.

3. Open access and state property regimes

According to Bromleys definition ownership and control over use rests in the hands of the state. Individuals can make use of the resources, but only at the forebearance of the state[17]. An open access regime is a situation where there is

---

[16] The Mouride brotherhood is one of the three Sufi Muslim orders which between them claim spiritual allegiance to the muslim population of Senegal. A particularity of the Mouride brotherhood is its involvuent in the peanut production and the organisation that the mouride saints have provided for a mass mouvement of agrarian settlement. Through a social, religious and economic organisation of pioneer farming communities (dara's), disciples of the mouride saints have been able to expand their agricultural activities to a vast, hitherto sparsely cultivated zone. (see D. Cruise O'Brien: "Saints and Politicians, essays in the organisation of a Senegalese peasant society").

[17] Bromley 1989 p.11.

no property, where the resources belong to the party to first exercise control of it[18].

As mentioned earlier it is difficult to distinguish between open access and state property in Senegal. Because of the 1964 land law, one regime tends to be the consequence of the other. Because of the limited resources to exercise its power, state ownership has in reality had little impact on the users of rangelands except for the important fact that it somehow inhibits establishment of efficient local management systems.

3.1 Co-habitation between agro-pastoralists and pastoralists in the Ferlo of Matam

In the previous examples we have mostly dealt with fairly sedentarised groups, although the existence of transhumant herding groups has been briefly mentioned. The following example from a non-project area outlines the problems of cohabitation between sedentarised agro-pastoralists and more mobile pastoral groups.

The study area, the eastern Ferlo of Matam, is inhabited/exploited by two major groups: 1) the original inhabitants (the "Ferlankes"), a group of agropastoralists with a fairly important agricultural production and a highly sedentarised lifestyle, and 2) the "Foutankes", originally agro-pastoralists from the Senegal river who have given up their agricultural production following the drought periods of 1973 and 1983/84 in order to save their herds.

On their move southwards the Foutanke herders adopted an extremely mobile lifestyle with no fixed point of attachment during the year, being in constant search of the best pasture lands for their animals. Following the years of good rainfall most of these foutankes (called "egge-egge's" meaning those who are always on the move) have however restricted their migration pattern to the vicinities of one or two deep-wells, in most cases combined with a limited agricultural production.

Contrary to the indigenous inhabitants, the Foutankes have concentrated on breeding sheep and goats. In a relatively short timespan after the drought[19]

---

[18] ibid. p.19.

[19] Unfortunately I do not dispose of reliable data neither of losses during the two latest drought periods nor on the degree of recovery.

they have managed to build up very large herds, many of them disposing of 700 to 1000 heads.

In order to secure good grazing especially for the youngest and weakest animals, the settlements are being moved every one or two months according to the availability of good pastures. The investment of labour in the pastoral activities is considerable as the small ruminants, contrary to the cattle, have to be accompagnied by one or several shepherds. Furthermore the settlements are usually located 10 to 15 km from the deep-well and large quantities of water have thus to be transported in huge rubber tubes on donkey carts to spare the youngest and weakest animals from the exhausting trip to the well. This labour investment is reflected in a much higher productivity in herds owned by Foutankes compared to those owned by their more sedentarized Ferlanke neighbours.

Inspired by the production results observed in the Foutanke herds, the Ferlanke agro-pastoralists, who formerly concentrated on cattle herding now tend to switch to small ruminants. Nonetheless, productivity in their herds remains much lower due to the constraints of their sedentarized life style and to the difficulties of interaction with the Foutanke production systems.

As the Ferlankes usually settle within a relatively limited radius around the deep-well, their animals graze in a ',centrifugal" movement, moving still farther away from the well as the dry season progresses. When the grass gets scarce their animals will therefore tend to reach areas already grazed by the Foutanke herds. The obvious result is that the Foutanke herds are in much better shape than the Ferlanke, a fact that has contributed to jealousy and hostility between the two groups.

The Ferlankes, having no means to expell other herders from the grazing lands have responded by trying to force the Foutankes to establish their camps in connection to the already established Ferlanke camps i.e. at the same "level" as the indigenous population or, even better, on their fields in order to exploit the manure. This is considered unacceptable to the foutankes, who claim that this model provides insufficient fodder for the young and less mobile animals.

In the village of Ranerou, the Ferlankes submitted the problem to the local authorities[20]. But as a consequence of the state ownership of the lands, these sided with the Foutankes and emphasized the equal rights of all citizens to settle wherever they want.

---

[20] The sous-prefet of Ourousougui.

The example thus confirms the observations of Stewe Lawry[21], stating that open access most commonly favours the richest segments among the herd owners, those who have sufficient cattle to give up agricultural subsistence production and adopt a nomadic life style. The state property regime does not provide any institutional framework to protect the grazing rights of the indigeneous population. At a first glimse it might therefore seem that the newcomers to the area, the Foutankes, have been the winners of the situation.

Nonetheless this might prove to be a superficial statement as the arrival of the Foutankes also has provided gains for the indigenous production system. An important contribution of the "strangers", recognized by the Ferlanke agro-pastoralists, is the innovations introduced in the raising of small ruminants. Furthermore sale of livestock from the Foutanke production units is usually undertaken by livestock dealers of Ferlanke origin, thereby enriching parts of the Ferlanke community. In general, strangers also tend to contribute more to the costs of operation of the wells than the indigenous populations who manage to diminish their watering taxes through personal arrangements with the members of the well committee.

Finally the presence of a large group of herders caracterized by a relatively low degree of self-sufficiency in cereals probably provides profitable opportunities of exchanges for the more surplus oriented Ferlanke agricultural production. In other words, the advantages and inconveniences of cohabitation between the two production systems should not be evaluated only through the competition of access to pastures. The introduction of exclusive rights is therefore not necessarily an adequate solution.

Discussion and conclusion

With the return to normal rainfall patterns, problems in the livestock sector have turned out to be less a problem of range degradation and overstocking than a question of competition between different production systems.

This competition has come more and more into prominence with the decline of previous forms of ethnical specialization. Where many semi-arid production systems previously rested on some degree of complementarity between different

---

[21] Lawry, S., 1990. "Tenure Policy towards Common Porperty Natural Resources in Sub-Saharan Africa", Land Tenure Center, Madison, Wisconsin.

ethnic groups[22], livestock is now owned by a large range of persons along a continuum from specialized animal husbandry to specialized farming. This inevitably leads to a more competitive approach to the available pastures.

The establishment of common property arrangements and/or the introduction of exclusive rights to a limited group of producers is, as illustrated by the examples presented in the paper, not always an appropriate solution.

First of all empirical studies have shown that the efficiency of pastoral systems relies on the maintenance of a high degree of mobility, incompatible with a rigid enforcement of exclusion rights. As shown in the case of the "Projet Sénégalo-Allemand" even minor calamities such as the breakdown of the deep-well or the passage of bush-fires require the possibilities of immidiate refuge into other areas.

Mobility remains one of the most important strategies of riskmanagement in pastoral societies. Interviws in the Ferlo confirms that a major factor contributing to the limited animal losses observed during the drought period of 83/84 was the promptness with which herders decided to move southwards once the failure of good rains was apparent[23]. The Foutankes have equally used mobility as a succesfull strategy of post-drought rehabilitation, limiting the lenght of their seasonal movements as the pastures regained their previous abundance.

This type of opportunistic strategies employed by the herders explains the reluctance of most pastoral groups to the introduction of exclusive rights. Exclusive rights are neccesarily related to the introduction of more conservative stocking rates. When the mobility of herds is limited, each animal tend to require more space for access to alternative foddersystems), as sufficient pasture must be available even in years of poor rainfall.

As stated by Sandford[24] a mildly conservative strategy would be based on a level of stocking for which sufficient pastures without overgrazing Would be

---

[22] Between for example Peul herders and wolof agriculturalist, (Touré, 0: 1986 "Peul du Ferlo", L'Harmattan, Paris) or maure camelherders and peul agropastoralist (Bonte and ould Cheikh 1982 "Production Pastorale et Production Marchande dans la Société Maure").

[23] Interviews with herders and functionaires of SODESP, 1988-90.

[24]Sandford, S. "Management of Pastoral Development in the Third World", John Wiley, 1983, cited from Livingstone, I: "The Common Property Problem and Pastoral Economic Behaviour" in "Agriculture and economic unstability", edited by Bellamy M.A. and Greenshield B. J. 0QEH. Aldershot, Gower, UK 1987.

available the 20 (possibly bad years) out of 100. Theoretically this signifies 80 years of excess of pastures i.e. relatively high opportunity costs.

Instead of accepting losses of potential production in many years most herders prefer high risk strategy where they benefit of high output and consumption during good years and meanwhile build up a large herd as insurance for future calamities.

These opportunistic land use strategies have proved to be rational under conditions of recurrent drought. They carry the implication of heavy losses - periodically and unpredictibly. But such losses can be dealt with in several ways, for example through insurance, diversification and spatial mobility. Even where common or private tenure regimes have been installed, for example in the "Projet Sénégalo-Allemand", building up of large herds remain the most important insurance strategy. This investment process is essential to understanding how range degradation or overstocking might occur under private or communal tenure systems as well as elsewhere.

A second problem related to the establisment of exclusive rights, is the fragility of the local institutions designated to enforce the rules. Local institutions will often lack authority to impose restrictions and sanctions even towards the indigenous group, particularly where there is little precedence for direct regulation.

As shown in the paper, only relatively loose institutional arrangements and restrictions existed for the management of natural resources before the colonial "intrusion" into the pastoral areas. As observed by Horowitz[25], very few pastoral societies are hierarchally organized. In most cases no individual has the authority to tell other members of the community how the latter should handle his animals. Many pastoral societies have no centralization of managerial decisions relating to access to grazing lands and water and therefore herdsize, composition, and movements.

The examples from the Bakel and PDESO livestock projects as well as the "Projet Sénégalo-Allemand" exhibit similar institutional weaknesses, leaving an important part of the rule-making and rule-enforcing to the project personnel.

The emergence of sufficiently strong support to enforce different kinds of rules and regulations will obviously be even more difficult when interests are heterogeneous. The conflicts of interest existing for example between Foutanke and Ferlanke agropastoralists are unliable to produce any institutional setting coherent enough to manage the natural resources in the area. In fact it is not absolutely evident that an interest of improving the management system, or even

---

[25]Horowitz, "The Sociology of Pastoralism and African Livestock Projects,AID Evaluation Discussion Paper No.6, 67 (1979), quoted from Lawry (1990).

the establishment of exclusion rights, is widely shared even among the indigenous inhabitants of the area. In years of relative abundance of pastures, gains as for example increases in collected watertaxes, terms of exchanges of cereals, or in the commercialisation of livestock might prove to be sufficient compensation for the inconveniences caused by the cohabitation.

With all its built-in conflicts the "open access regimes", functionning in most of the Ferlo, may turn out to be more flexible than property regimes relying on the introduction of exclusive rights. With all its apparent shortcommings (best illustrated by the chaotic management of the deep-wells), this system permits the cohabitation of different production systems each with their comparative advantages[26].

The inherent assumption of most interventions in the livestock sector is that more control is neccessary. Such control can very well prove impossible, given the almost constant flux as well as the still evolving social relations between the different groups. Regulations or associations that presumably are accepted by the pastoralist in rain deficit periods may well break down in periods where rainfall is more abundant, permitting pastoralists to return to less structured grazing patterns. As stated by Bennet et al, "all management plans must be extremely flexible and take into account the differences between the different groups. Nothing seems to indicate that there is only one right system for everybody".

---

[26]In fact ongoing field research at the Centre de Suivi ycologigue indicates, that despite of the open access regime, the spacial distribution of different types of production is more structured than formerly presumed.

Water resource management

Jean-Pierre Zafiryadis

1. Availability and exploration of water resources

The availability of water resources has always been the key constraint to development in the Sahelian Zone. During the last decade the exploitation of the water resources has increased dramatically. There are three elements which have a cumulative negative effect on the water resources:

- Eratic and insufficient rainfall

- Increase of the population of the Sahelian Zone, leading to increased exploitation of water resources

- Introduction and development of new and highly water demanding activities such as large scale irrigation, industrial activities and a large increase in water demand from urban consumers.

The increasing exploitation of water resources is not only a quantitative problem, but also to a large extent a water quality and pollution problem.

Among the effects of this development is an increasing conflict of interest in connection with the exploitation of water resources. Consequently, there is a growing need for formulation and implementation of a rational and updated water resource planning and management policy.

The water resources in the Sahelian Zone can very roughly be divided into four categories:

Major rivers. There are seven major rivers passing through drought prone areas (the Senegal, Niger, Nil, Tana, Athi, Shevelli, and Juba Rivers). These major rivers are vital as sources of drinking water for towns in the Sahelian Zone, which are mostly located on these rivers. Therefore, pro-tection against pollution and overexploitation is necessary in order to ensure an adequate drinking water supply for large population concentrations in the Sahelian Area. Furthermore, these rivers have a significant potential for irrigation which could be carefully developed. But the management of these resources entails problems of transboundary water resources among various purposes.

Other surface water resources. Apart from major rivers the Sahelian Zone contains a large number of other surface water resources, which generally dry out during various periods of the year. Therefore, these resources can in many cases not be used directly as a source of drinking water. They can however, be used for agricultural activities, but there is of course a need for proper management of this resources, since they can represent health hazard due to contamination. Inadequate development can lead to the loss of the water resource for large groups of populations due to increased evaporation and pollution of the water resources.

Ground Water resources. The well is a symbol in the Sahelian Zone. Ground water is extensively exploited for human and livestock consumption. During the last decade an impressive effort has been made to increase the ground water based water supply by the drilling of thousands of wells, most of them equipped with hand pumps, and other low-yield extraction for drinking purposes. Shallow wells and boreholes are also used to a limited extent for supplementary irrigation purposes. Due to the very slow recharge of the ground water reservoirs, the abstraction of ground water should always take place on a sustainable basis, which means that the abstraction rate should not exceed the safe yield of the aquifer. Otherwise the consequence can be a dramatic lowering of the water table, or even a risk of deterioration in water quality, thus threatening the very existence of large populations.

Very large Ground Water Resources. The case of very large aquifers deserves separate mention such as for example the deep aquifer of the Tchad Lake Basin. Some of these aquifers have an excellent yield, and in some cases they are even artesian, therefore permitting extensive exploitation for irrigation. The attention has nevertheless to be drawn to the water quality of some of these aquifers, where the water has a certain level of mineralisation, which can rather quickly lead to salinisation of the soil in irrigated areas. Even if it is acceptable to extract a larger amount of groundwater than its recharge allows for a period of time, considering the water resource as a defined definite resource which can be exploited as coal and oil, where an extensive exploitation of the resource can eventually lead to a lowering of the water table and constitute a threat to the drinking water supply for the populations of the area.

2. The Constraints within Water Resource Management

Technical constraints

As in most of the developing countries the technologies goes from very simple traditional technologies to the most sophisticated within the same field. Furthermore, the Sahelian Zone is generally very low populated and the distances are important. Under these conditions the disponibility of spare parts, know-how and back-up for an impressive variety of equipments is very restricted.

A lot of work has been done in the field of technical adaptation of the technologies involved in the water management and development. But in many cases these methods and technologies have not been sufficiently adapted to local conditions, both in terms of technology and in terms of affordability. Many examples can be given, e.g. monitoring systems which are too sophisticated to be reliable under Sahelian conditions, ground water pumping technique, which can not be maintained properly under local conditions. There is therefore still a need for further improvement in the technologies involved in water management.

Economic constraints

The economic environment of the water resource management is characterised by two major constraints.
- The Sahelian countries belong to the poorest group of countries in the world.
- The physical conditions existing in the Sahelian zone make cost levels higher than in other parts of the world, e.g. the high cost of transportation due to very long distances and poor roads, combined with high fuel prices, the scarcity of local managers and trained technicians resulting in inadequate implementation or costly expatriates, and the limited availability of mechanical equipment and inadequate supply of spare parts.

As a consequence of these two major constraints the proper long term resource management demands a considerable political willingness and a capability to deal with the various conflicts of interest. Furthermore the poor own capacity of investment make to a large extent these countries very dependent of un-coordinated donors assistance which is not always consistent with the country's own policy.

Institutional constraints

In situations where the technology is not always well adapted to local conditions and economic conditions are very difficult, demands on the institutions to ensure proper water management are very high. But, however, the existing institutions are not always the most appropriate, and the capacity of the governments to extend and improve public services in general is very limited, due to the conditions of the national economies.

Before adressing the situation in the Sahelian Zone it should be mentioned that institutional and human resource development means different things in different contexts to different people.

Institutional development is the process of improving an institution's ability to make effective use of the human and financial resources available.

This process can be internally generated by managers, staff or members of an institution or induced and promoted by a government or by development agencies.

The broad concept of institution encompasses entities at the local or community level, project management units, parastatals, line agencies in the central government, etc.

Tasks included in institutional development are management systems, including monitoring and evaluation; organisational structure and changes; formulation of sectorial policies, planning and interagency coordination; staffing and personnel policies; staff training; financial performance, including financial management and planning, budgeting, accounting, and auditing; maintenance; and procurement.

Internationally it is today recognized that there has been too much emphasis on the economic and technical aspects of the development process, while the institutional and human resource development aspects have not received adequate attention, considering the potential for long term environmental cum economic returns on investments in this sector.

The institutions in the Sahelian Zone, where the level of education is among the lowest in the world, and where the institutions and the organisational thinking is still influenced by the colonial heritage, are not in general designed to meet the management needs of an independent country in the 90-ties.

This is fully reflected in the water sector, where the institutions involved are generally very weak, being only recently established as planning ministries or ministries of environment and have thereby only been able to command a small share of the necessary resources from a human resource point of view. Most of the institutions need adequate staffing, both in terms of numbers and skills. In a period of continuous economical crisis in most African countries, this situation makes it very difficult for these institutions to compete with well-established institutions within the fields of agriculture, trade, industry or urban development.

Also the regulatory framework (legislation, including control and enforcement rules) within which the institutions function is in general not updated to meet the needs of the present situation.

2. The Water Resource Management

Proper water resource management should be based on three main principles:

Strict and clear priority for the utilisation of water. This priority should give a priority to human and animal consumption and traditional methods of cultivation.

Policies defining priorities for the use of water resources require adequate knowledge of the available water resources and their possible development and an accurate assessment of the water resource requirements for different uses. But it also requires an appropriate institutional framework which can define and enforce the policies and priorities for the exploitation of water resources.

Ensure efficient and adequate use of the water resources. An increase of the efficiency of the water use should be developed within all fields of utilisation of the water resource.

Improving the efficiency of water utilisation will generally require the introduction of new techniques or the modification of existing techniques, e.g. better irrigation techniques, adequate choice of crops, industrial activities with low water demands, water saving measures, etc. Obviously these developments require added investments. There are furthermore very heavy demands on the institutions, which are responsible for the existence of adequate laws and law enforcement, sensibi-lisation, adequate incentives, etc.

Ensure adequate development and conservation of the water resources. Improvements in the protection of water resources involves the introduction of new drilling techniques, pumping techniques and water treatment techniques, among others. The development of water resources is generally an active inter-ference in the nature, and will therefore involve secondary effects at short term and at long term, which has to be dealt with before implementing any development activities. The water resource development involves generally heavy costs and demand imperatively the active involvement of institutions at all levels, in order to ensure the good use of the investments.

Management of natural forests by local people in Burkina Faso, Sudano-Sahel region, West Africa

Per Christensen

The overwhelming ecological problem in Burkina Faso (and in a lot of other Sahel-countries) is the ongoing desertification, the overexploitation of the environment/nature. This is due to overpopulation, unsustainable cattleherding, unsustainable agriculture, bushfires and a large consumption of fuelwood.

For foresters it has for a long time seemed logical to fight desertification and overexploitation by plantations. This is still true in many countries, but I think that we in this way neglect the most important principle in all first aid, which is:

> First: Stop the accident.
> Then: Lifesaving first aid.

By producing vast areas of plantations we have not stopped the accident. We are just trying to heal the wounds of the accident, but at the same time the "accident" just carries on. While we are healing wounds in one area, new wounds are inflicted to the environment in other areas. In our case more clearings are made, and more forests are cut down or in other ways distroyed.

In other words, the question of preventive actions contra curative actions.

A MANAGEMENT MODEL DEVELOPED FOR A NATURAL FOREST IN BURKINA FASO

It started out as a pure FAO forestry project in the Nazinon area, but as it was a pilot-project, it has changed set-up and tactics many times to become what it is today.

What will be described here is what you would call the ideal setup so far. Don't get the impression that the project got to this point right away, but anyway, this is where it is today:

To secure ecological sustainable use of the nature, it must also be economical sustainable - at least seen from the local peoples point of view. They can't afford to be motivated just from ecological aspects. No doubt about that they know what is ecological sustainable, but they have to feed their family tomorrow also. So if they don't see an immediate benefit in what you would like them to participate in, they will not do it.

This is partly why this forestry project choose to intervene in this rather southern area. Here is a forest <u>potential</u> which is sufficiently interesting also from the farmers economic point of view.

So to summarise: We have gone so far south, that the "accident" has not yet arrived, which means we can <u>prevent</u> the accident. And furthermore this area has a potential that makes it economical interesting as well as ecological.

The method used by the FAO Project to put up a management model for nature forests in the North-soudano region comprise:

- Definition of management objectives.
- Quantification of human and forest ressources,
- Characterization of the wood market,
- Identification of restraints,
- Elaboration of appropiate technical and sociological instruments.

The objectives of the management of natural forests have been defined relative to the forest and the tenure rights in place, the legislation concerning the association of the farmers in cooperatives and the national policy on management matters formulated in 1981. These objectives are the following:

- Preserve the natural forest stands by appreciation of their value, with the active participation of the rural population.

- Raise the production and the productivity of the natural forests, by red foresting the cleared areas and by enriching the degraded areas.

- Contribute to the assurance of fuel-wood self suffiency for the rural and the urban population by creating sources of sustained production by rational exploitation of the natural forest cover.

- Contribute to the economic and social development of the villages associated with the management of the natural forests, by creating rural employment, - organizing cooperatives and monetarize the income.

<u>The components of the model</u>

The model developed by the project carries three components:

-   The <u>technical</u> component is responsible for the preservation of the sustained production of the natural forests.

-   The <u>socio-economic</u> component is responsible for the identification and the development of the different forms of participation of the rural population:

-   The <u>educational</u> component is responsible for the long term assurance of sustainable management models in natural forests.

The criteria and the activities of each component are mentioned in the following:

THE TECHNICAL COMPONENT

1. Identification and reconnaissance

<u>Cartographic analysis</u>: A preliminary analysis of the disponibility and the localization of the forest resources, and of the population in the proximity of the forest, characteristics of the road network and the distances to the consumer centers and, by consequence, the viability of the possibilities of synthesizing these potentials into a management plan, is carried out on the basis of the available cartographic informations in scale 1:200.000. This analysis permits the identification of the forest zones fit for management.

    <u>Verification of the potential</u>: The verification of the forest potential identified in scale 1:200.000, should be done on the basis of recent satelite images, eventually combined with a systematic reconnaissance of the forest stands in order to identify the modifications experienced between the date of the images and today.

    <u>Aerial and land reconnaissance:</u> Aerial and land reconnaissance protocoles of the forest stands are worked out in order to reunite the basic data necessary to make the final choice of zones to put under management.

    <u>Information of the authorities concerned.</u> Once the choice of which areas to put under management is made, it is absolutely necessary to inform the central, regional and local authorities concerned by the management, in order to avoid any contradictions with other programmes or projects (planned or already working) in the area.

2. Aerial photographs and cartography

Aerial photographs: Recent aerial photographs are indispensable for the elaboration of an updated-cartographic basis. Experience has shown that the scale 1:20.000 and the period just after the rains and before the bush-fires are the most appropriate.

Cartographic inventory and land use: The photo-interpretatation and the cartographic work should be oriented towards the land use. The principal factors are:

The different types of forest stands

> The human presence/land use
> The hydrography
> The road network

3. Identification of management units

Definition of management Units: The use of the Management Unit concept is necessary in order to give specific physical coordinates to the administrative responsibility of the cooperatives in charge of the execution of the management plan, and to the Forest Service in charge of the assesment of the management plan.

A Management Unit is defined as a surface of between 2.000 and 4.000 ha composed by a number of parcelles equivalent to the adapted rotation. The dimensions of the Management Unit are determined by socio-economic and ecologic criteria. The lower limit (2.000 ha) is guided by the minimal annual production necessary in order to keep the interest of the cooperatives. The upper limit (4.000 ha) is guided by the concern of not cutting more than 200 ha (rotation 20 years) of continuous area per year, and of not forcing too many villages to work together in order to obtain the necessary number of workers to carry out the management plan.

Identification of Management Units: This identification should be base on the cartographic inventory of land use, the localisation of the neighbouring villages and their affinities, the number of villagers interested in the management and finally the disposition of primary and secondary access roads.

With this data, it is possible to formulate a first draft identification of the Management Units, which should be submitted for sanction to the cooperatives before application.

The materialization of the Management Unit limits in the field takes, whenever possible, the natural limits (waterways, roads, relief etc.) in use. These are completed, if necessary, by the installation of permanent roads/firebreaks.

4. The installation of parcelles

The definition of parcelles: The parcel is the unit of annual intervention in the forest. Its surface extent is determined by the total surface of the unit divided by the number of years within the rotation according to the management plan.

The demarkation of the parcelles: The delimitation is first done on the map, and afterwards carried out in the field. Here, whenever possible, natural limits are used, and these are completed, if necessary, by metal signs or boundary posts, but no permanent roads/firebreaks are installed.

5. Appraisal of potentials and constraints

The method of appraisal and its objectives: The method of appraisal of potentials and constraints to management is one of the tools worked out by the project in order to measure and assess all the parameters characterising the natural forests in the Sudano-sahelian region in their dynamic context.

Its objectives are:

- Assessment of the standing volume of fuel-wood in each parcelle.
- Appraisal of the ecological situation of each parcelle.
- Verification of the degree of human activities and the identification of probable solutions to the conflicts of the interests of the local population and the objectives of the management plan.
- Localization of eroded areas and determination of the causes and the degree of the erosion.
- Verification of the accessibility of the parcelles by lorries in order to secure the commercialisation.
- Verification of the photo-interpretation.

The appraisal: The appraisal work is executed after the materialization of the parcelles in the field. This work is followed by a treatment of the data in order to obtain a final descriptive sheet of each parcelle and a general summary of the whole Management Unit.

6. Elaboration of the management plan

The Management Plan is a document which synthesizes the basic data concerning cartography, sociology, economy and forestry in order to present and justify the best intervention alternatives available for reaching the management objectives.

These interventions concern mainly; rotation, cutting regime, sylvicultural methods, protection methods and other activities such as agriculture and pastoralism.

The technical part of the Management Plan deals with the specific interventions in each parcelle throughout the rotation period. This part of the Management Plan should be formulated in a language that makes it comprehensible to the members of the cooperatives, responsible for the execution, and to the forester in the field, responsible for monitoring this execution.

The Management Plan (with regular revisions) should secure a self-supporting execution of all the necessary interventions during the whole rotation.

THE SOCIO-ECONOMIC COMPONENT

1. Identification of the affected villages: On the basis of the cartographic analysis, a preliminary identification and a first round of contacts is made with the villages adjoining the forest.

At this first meeting the village authorities are consulted concerning:

- The programmed activities of the project/Forest Service in order to appreciate the natural forest.

- The localisation of their village terretories relative to the management zone.

- Their willingness to participate in the management of the forest.

If the village authorities declare the village ready to participate in the management of the forest, a second meeting is set up with all the interested villagers, who are convoked by the village authorities and not by the project/ Forest Service.

2. Village extension: This second series of meetings is oriented towards a deepening of the knowledge of the village history and of the villagers concern for and knowledge of the deterioration of the vegetation cover and the degradation of the soil. These meetings, open for all interested, are held as open dialogues with the active participation of the villagers, often split up into groups of women, young people and men. At the end, the villagers interested in the management of the forest, are asked to form a cooperative and to set up a third meeting in order to clarify the details concerning the management and the active participation of the cooperative.

3. The Forest Cooperatives: The third series of meetings should materialize in the constitution of the cooperatives, register the names of the management committee and the number of villagers ready to participate in the management.

The planned activities (exploitation, silviculture and protection) should be clearly explained, and it should be stressed that these forest activities are complementary to the agricultural activities. And this is what is to be the sustainable part of the project.

-   The forest exploitation takes place after the harvest until the beginning of the preparation of the fields (January - April)

-   The silvicultural work is limited to a maximum of two weeks in the beginning of the rainy season (June - July). This should not interfere too much with the agricultural seeding.

-   The protection of the forest against bush/forest-fires (preparation of firebreaks and application of early controlled fires) takes place just before the harvest period (October). This is normally the period where the farmers have no more cereals in their stock, and a monetary income will permit them to buy the necessary food for the family.

As a consequence, a strictly forest cooperative does not exist, and the cooperative members should guard their status as farmers.

Finally the cooperative is asked to designate one person (monitor) per 20 members, who will receive education in the elementary techniques of forest management and who should afterwards transmit his knowledge to the rest of the members.

4. The Forest Management Fund has been created in agreement with the cooperatives and the Forest Service. Its main objective is to create a means to ensure the execution and the assessment of the management plan during the entire rotation.

The financial source for this fund is the fuelwood sale. The official price per stere is 1610 F CFA ($\approx$ 5,4 \$), and this sum is divided along the following lines:

Woodcutters direct salary	610F CFA
Revolving account of the cooperative	200F CFA
Forest tax	300F CFA
Forest Management Fund	500F CFA
	1.610F CFA

How the Forest Management Fund works: The management committee of the cooperative is responsible for the registration of the entire fuelwood production of all its members. Furthermore this committee collects and distributes the woodcutters salaries, and it collects and administers the Revolving Fund af the cooperative.

The Sales Clerk (payed by the Forest Management Fund) conducts the wholesale fuelwood merchants towards the stocks. controls the loads and delivers receipts. The Sales Clerk is responsable of paying the cooperative committees the 810 F CFA, and he sends the remaining 800 F CFA to the central authorities, who transfer the Forest Tax (300 F CFA) to the Treasury and keep the 500 F CFA in the Forest Management Fund.

For the time being, the Forest Management Fund is administered by the Project/Forest Service.

THE EDUCATION COMPONENT

The foresters associated with the project receives consecutively on-the-job training in all areas of management of natural forests. - Both theoretical and field training.

Some internal courses of short duration within the project and study tours to similar projects are also carried out.

The Sales Clerks are locally recruited personnel who will be in charge of the entire process of the commerce of the fuelwood. These Clerks receive a short education in the areas of forest legislation, administration, commerce and accounts.

The Forest Management Monitors receive an education of 10 weeks (paid by the Management Fund.

This is both theoretical, but especially practical training in the different management activities. The training takes place in the cooperatives "own" forest at the different appropiate periods. Each year a certain number of already trained farmers participate in the neighbouring areas training.

6 weeks in the area of exploitation in January-February.
2 weeks in the area of silviculture in June-July.
2 weeks in the area of protection in October.

TECHNIQUES

Shortly about the different techniques:

Cutting:
First dead wood is cut in the area which is to be harvested this year (rotation 20 years). Dead wood can be sold immediately, and give quick money/motivation.

After this, about 50% of the live trees are marked for cutting after certain criteria. This marking is done by the farmers/woodcutters themselves (after training). These trees are cut and stacked for later sale.

Sowing:
Most trees sprout from the stump, but in addition to this seeds from local species are treated and sowed by the members of the cooperatives (after training). The local people have themselves chosen the species they are interested in.

The areas sown are the "naked/eroded" areas in the forest and the "poorer" areas (related to species). Wildings can also be used. (Parallel trials are being done by the project with direct sowing, wildings, grafting etc.). No nurseries are put up for this activity due to the vast areas treated (2000 ha). All work is being paid by the 500 F CFA (Fonds d'Amenagement).

Protection:
Firebreaks of 3 m's large are installed round the areas to protect. These are the areas newly harvested and sowed (protected for 5 years) The rest of the forest is burned with early controlled fires by the local people under the supervision of foresters.

This works as a double protection

- against the fires, and
- against grassing of the new sprouts and seedlings.

In the areas burnt by early fircs, the cattle is allowed to browse the new fresh sprouts of grass provoked by the fire. In the non-burnt areas the high, dry uninteresting grass protects the new seedlings. All work is paid by the 500 F CFA (Management Fund).

Harvest:
Fruits and seeds are harvested in the forest by the local people throughout the season. These seeds are collected and stored in a central place. Paid per kilo by the Management Fund.

RESULTS AND THE FUTURE

Little by little other activities are introduced into this forestry project, like for example bee-keeping, improved stoves, fruit trees, animal traction, haymaking and so one. These "secondary" activities induce a more even distribution of the benefits. Women, men and children can participate on almost equal terms. Herders have been integrated in the forest management, and young people find occupation in their home-village, which means they do not have to emigrate in the dead periods of agriculture.

Rural population has got a source of monetary income, which can give them access to credit possibilities and further development.
The money from cooperative fund (200 F CFA) are used for general purposes for the benefit of the whole village - not just for the forest cooperative.

THE NAZINON EXPERIENCE - SOME NUMBERS

Very briefly the preliminary results of the application of the management model, described below, on 25.000 ha of natural forest in the Nazinon area in Burkina Faso are:

Rotation:	20 years
Management Units:	7 Units with a total number of 20 villages participating.
Cutting regime:	Approximately 50% of the standing volume is cut according to specified criteria
Woodcutters:	From 20 villages a total of approximately 600 woodcutters are participating in the management.
Forest Management Fund:	Operational since 1988. In 1989 the turnover was 12.000.000 F CFA ($\approx$ 40.000 $)

Unso's Experience from Projects/Programmes in the Sudano-Sahel[27]

Henrik S. Marcussen

Abstract

UNSO has over the years established a long range of projects and programmes in its 22 mandate countries in the Sudano-Sahelian region. The value of UNSO's actual project portfolio is around US$ 100 million, and the annual turnover around US$ 30 million.

UNS0 has constantly tried to learn from past experiences when extending projects into new phases or formulating new projects. Certain achievements, especially on the technical side, have proved invaluable for new projects, not only UNSO's, but other agencies' projects as well. Shortcomings have been related particularly to socio-economic issues, where the involvement of the local population in project design and implementation has proved crucial for project viability and
sustainability.

While maintaining a certain "project nucleus", a more integrated approach to natural resource management projects is needed, in order to address simultaneously several of the factors causing ecological degradation. Other lessons learned have been that smaller scale projects, which are more easily manageable and have more scope for replication, are preferable; indigenous technologies (and species) should be further researched, utilized and exploited, in cases combined with imported, adapted technologies; project results should be of immediate material benefit to the local population, if projects are to succeed; the protection/rehabilitation of the productive natural resource base should be given high priority to enable the vicious circle of poverty resulting from, and at the same time causing degradation, to be broken; a more flexible approach to project design should be adopted, by operating with a "menu of options" type of strategy. And, most importantly, popular participation should be ensured.

---

[27]Paper delivered to the IUCN General Assembly Workshop on Conservation and Sustainable Development in the Sahel and Other Arid Regions, Perth, Australia, 28th of November - 5th of December 1990.

UNSO : What it is, what it does

UNSO (The United Nations Sudano-Sahelian Office) was established in 1973, following the prolonged drought that struck the western Sahel during 1968 - 1973. UNSO was then entrusted with the responsibility of assisting the countries afflicted by the drought in the Sahel in their medium- and long-term recovery and rehabilitation programme.

Later, in 1976, UNSO was transferred from the SecretaryGeneral's office to UNDP, and in 1978 its mandate was extended to include the responsibility of assisting 22 countries in the Sudano-Sahelian region to implement the Plan of Action to Combat Desertification (adopted in 1977), on behalf of the United Nations Environment Programme (UNEP)

UNSO's main activities are the following:

- Co-operating with Governments in planning, co-ordination and ecological monitoring;

- Implementing Projects and Programmes;

- Mobilizing financial resources;

- Facilitating international co-ordination;

- Serving as the United Nations focal point for regional organizations, in particular the Permanent Interstate Committee on Drought Control in the Sahel (CILSS) and the Intergovernmental Authority on Drought and Development (IGADD);

- Providing information and awareness creation.

Over the years, UNSO has established a long range of projects within deforestation control, rangeland management, soil and water conservation, dune stabilization, and integrated land management projects, etc. The value of UNSO's actual project portfolio is around US$ 100 million, and its annual turnover is approximately US$ 30 million".

Compared to the size of its project portfolio - and in other respects as well - UNSO is a small organization, with only 18 staff members at the professional level at Headquarters, most of whom are in the Programme Division (eight of them deal, on a day-to-day basis, with all administrative and management problems related to projects and programmes). The Technical Support Division has five staff members at professional grade who provide technical backstopping, give advice, develop project ideas and concepts and write position papers, etc.

In addition, UNSO has two small regional offices, one in Nairobi, Kenya liaising with UNEP and IGADD, and one in Ouagadougou, Burkina Faso liaising with CILSS. Each regional office is staffed by two professional grade staff members.

For the smooth running of its projects and programmes, UNSO relies heavily on local UNDP offices. To ensure an efficient local administration, UNSO employs national officers in all major programme countries or, in a few cases, a JPO, who take care of UNSO matters within the UNDP offices.

Experiences from operational projects and programmes

As mentioned, UNSO has had a vast range of projects over the years. The projects have provided - in spite of some of their shortcomings - a useful and important technical insight, which has proved indispensable when designing and formulating new projects, and have enabled us to learn from past experience. In the following, two types of projects will be presented with a view to demonstrating how UNSO has tried to adopt new approaches to project formulation and design, continuously trying to take past experiences into consideration.

Towards the end of this section, the presentation of the two project types will be used for formulating some basic principles on which future projects -to our mind - will have to be built.

a) Dune stabilization projects

UNS0 has carried out a number of sand dune stabilization projects, in Senegal, and - on a somewhat larger scale - in Somalia and Mauritania.

Based on the experiences gained from an UNSO-financed project during the period 1981-1983, which studied the desertification problem in Mauritania and then formulated a national strategy, a dune stabilization project was formulated in 1982 and implemented during 1983-1986. UNSO's participation in the project was ensured by a financial contribution of approximately US$ 3 million from DANIDA, while contributions from UNCDF, UNDP and WFP together came to another US$ 3 million. Furthermore, with relation to the Dune stabilization project, UNSO, with a contribution from DANIDA, has financed a project aimed at strengthening national institutional capacities within planning, ressource assessment and monitoring.

The objectives of the project in its first experimental phase were mainly:

- to conduct experiments and trials at the research station at Boutilimit on developing species best adapted for biological stabilization of dunes;

- to conduct experiments on various stabilization techniques at 15 different locations, selected as being representative of different conditions for dune formation; and

- to develop a methodology for the diffusion and integration of the experience and knowledge gained, enabling the local population to increasingly use these stabilization techniques, at their own initiative, in protecting their productive resources.

In this first phase, the project experimented with various techniques, combining mechanical and biological stabilization techniques, on 15 different project sites along the main infrastructural link in Mauritania, the Route de l'Espoir ("The Road of Good Hope"), which is constanly threatened by sand and dune movements.

The experiments at the research station, and also in the field, indicated that one exotic species (Prosopis juliflora) and one local species (Leptadenia pyrotechnica) were particularly promising, even under severe growth conditions with very limited amounts of rainfall.

The main results of the first phase could be summarized as follows: it proved that it was technically possible to stabilize dunes under various ecological conditions, even in locations with rainfall between 150 and 200 mm, and both planting techniques and species selection were extremely promising. Coping with the global problem of sand and dune movements in Mauritania would, however, be excluded for a number of reasons. In the context of the project it became obvious that the costs involved, using the combined mechanical and biological stabilization techniques, were prohibitive for any large-scale effort at stabilizing dunes (the costs were estimated at close to US$ 1000 per ha.).

A major shift in project design and approach thus took place when formulating the second phase of the project (1987-1990), turning the emphasis from curative to preventive measures, and emphasizing the protection/rehabilitation of farmers' and herders' productive resources, rather than the protection of the physical infrastructure which, however useful and necessary in certain contexts, proved to be extremely difficult. No popular participation could be mobilized for the maintenance of stabilized areas along the roadside, and the project's sustainability was thus clearly at risk.

The project approach in this phase was thus turned into a much more integrated approach, combining developed stabilization techniques, viz. farmers' and herders' productive resource base, with efforts in improving productivity and income opportunities by soil and water conservation techniques, agro-forestry plantings, etc. At the same time, much more emphasis was given to the participatory issue by consulting farmers and herders as to where stabilization could most effectively be done, and by the project offering transportation of material for stabilization purposes for individual or group-organized efforts in protecting individual or village/community lots, etc.

In other words, the first phase revealed obvious shortcomings, while at the same time certain necessary techniques were developed and tested, indispensable for later interventions. The second phase attempted to learn a lesson, emphasizing preventive rather than curative measures, recognizing that the phenomenon of ecological degradation/desertification is a complex issue requiring various simultaneous project interventions and activities dealing with several of the causes of degradation, thus advocating the integrated approach, and seeking to protect/rehabilitate productive resources, rather than infrastructure, for better future management of resources and improved living conditions.

The most important, and generally applicable lesson, was certainly that public participation is essential for project viability and sustainability, and that such participation can be achieved mainly to the extent that direct producers will have an immediate material gain from project activities.

b)   **Fuelwood plantation projects**

The fuelwood plantations in <u>Ethiopia</u> in many ways illustrate points similar to the ones raised above. In addition, however, they point to severe constraints in reaching objectives, some of which are inherent in the project design, others mainly due to lack of incentives for the direct producers, closely associated with past and prevailing Government policies.

The first UNSO-supported fuelwood plantations in Ethiopia were started in 1984 in Debre Birhan and Nazret (with DANIDA financing) and in Dese (with financing from the Finnish Government). Later on projects were formulated for

Gonder and Hararge, all of which tried to incorporate the lessons learned from the earlier projects.

Recognizing the very serious degradation in the highlands of Ethiopia, following extended cultivation, overgrazing and cutting of the vegetation for fuelwood needs, the aim of the projects was to help to arrest and reverse the process of deforestation and to contribute to the protection and enhancement of the Ethiopian environment needed to sustain the productivity of the country's natural resources.

In 1990 a total of 6,000 ha in Nazret and 3,600 ha in Debre Birhan had been planted, mainly with Eucalyptus globulus in block-type plantations which, at the time of project conception were believed to be the most economical and efficient to establish and manage, and which were expected to protect areas downstream of the project sites by full watershed management. The exotic species, Ecalyptus globulus, introduced already in 1895 by Emperor Menelik II, was selected for planting because of its performance in terms of survival, growth and quality of the fuelwood produced, this exotic species was selected for planting.

In more detail, the objectives of the projects were:

-   to provide fuelwood for the number of residents anticipated in the towns of Nazret and Debre Birhan in 1992;

-   to strengthen the Ethiopian institutional capacity for planning and managing fuelwood projects;

-   to assist in the development of community relations and to provide social services;

-   to create employment in project areas;

-   to develop soil and water conservation measures, and improve soil conditions;

-   to provide fuel and other forest products also to residents of the area itself, in order to reduce pressure on remaining resources.

Despite highly irregular rainfall, and the fact that in the higher project altitudes prevailing at the Debre Birhan site - there are more than 15 days of frost annually - the projects have fared well technically. Fuelwood production is expected to be close to target and a sustainable and easily manageable forest, producing a very large economic surplus given current official - not to mention unofficial - prices is in place.

Apart from this, well functioning nurseries are installed, construction of a network of secondary roads has taken place, housing, storage facilities, schools and health clinics are established, experiments with agroforestry and soil and water conservation techniques have been carried out in co-operation with the farmers, and a great number of people have been employed in the nurseries, in the planting of seedlings and the construction of roads, etc.

Nevertheless, UNSO does not intend to repeat this kind of project design, and has tried to modify the approach throughout when taking the earlier projects into new phases, and when formulating new projects intended to meet the fuelwood needs of the population, whether in urban or rural areas.

The conceptual flaws of the projects originally designed are not difficult to identify. Block type plantations, with plantings on land formerly used for grazing or farming purposes, however degraded and eroded should, in principle, not take place. Restoration of degraded land and soil structures and the production of the necessary fuelwood should take place with the farmers and direct producers, by using agroforestry techniques and soil and water conservation measures, even if that would imply slower growth of trees and less output of fuelwood. Creating virtually monocultural plantations with exotic species is another obvious risk.

The major problem, however, is that the local population has not been fully involved in project implementation, except when hired as casual labourers which, of course, has nothing to do with the concept of participation; and to date there is no guarantee that the farmers and herders in the project area will share in the benefits derived from profect activities, for instance by being granted the right to harvest and sell fuelwood produce, under the supervision of project technicians. UNSO has been arguing in favour of a marketing, selling and pricing structure which would benefit the local population while at the same time ensuring that the Government receives a considerable return on its investment; but so far, this has not been achieved.

This is even more critical because the peasants and herders in the project areas have had to give up some land for planting purposes when the project was established. If the local population only has to bear some of the costs, without reaping part of the benefits derived from project activities, conflicts can easily be imagined, jeopardizing project results.

As mentioned, UNSO would probably not enter into another project providing fuelwood based on large, block-type plantations (not to mention plantations of a monocultural type). Rather, we have tried to encourage new concepts with regard to project designs in Ethiopia, whereby fuelwood needs will be met by agroforestry and community forestry production, and where projects have a much more integrated approach, seeking to improve on agricultural productivity while at the same time trying to establish a more balanced utilization of pastures, introducing soil and water conservation techniques, etc. The main principle in this design is, however, to ensure public participation by allowing farmers and herders to influence project implementation but, in particular, to share in the benefits derived from project activities.

The lessons learned from these fuelwood plantation projects in Ethiopia resemble in many ways those described above in connection with the sand dune stabilization projects: the need for a more integrated approach, where elements in the projects address several issues related to the process of ecological degradation simultaneously, establish solutions to ecological degradation for and with the direct producers, meeting their immediate demands and improving on living conditions by introducing techniques which are easily manageable and affordable, etc., etc.

However, our experience in Ethiopia illustrates one aspect in particular which, to a greater or lesser degree, is also characteristic of many other countries in the Sudano-Sahelian region. The project designs and approaches discussed above may have been inappropriate and in some respects, even detrimental to the efforts made in establishing viable, sustainable and ecologically sound projects. But even with the introduction of corrective measures and more appropriate project solutions formulated, prevailing Government policies can be a major adverse factor. In the case of Ethiopia, villagization policies and quota systems, obliging farmers to sell certain portions of their output to state bodies at fixed prices, the lack of appropriate rules and regulations pertaining to access to land and user/property rights, and other regulations affecting the individual producer's access to markets etc. etc., are all elements which can negatively influence efforts to restore the ecological balance.

c)   Newly developed UNSO projects

Even with an integrated approach, the concept of the project might not be feasible in a greater context, as it is bound to be a punctual intervention with somewhat limited impact. An alternative to this is a programme approach, recently adopted by UNSO in collaboration with SIDA.

Financed by SIDA, UNSO has established a number of quite similar project types in Senegal, Burkina Faso and Niger (5 projects in all at a present value of approximately US$ 16 million). As these are all integrated resource management projects, the emphasis has been on the introduction of agroforestry techniques, soil and water conservation measures, techniques for more sustainable management of the resource base (in particular water and pastures), mixed farming practices, etc.

The approach is not to present the project as a fait accompli, but to operate with a relatively long introductory period (up to 2 years), allowing for a socio-economic baseline and other surveys, thorough consultations with the local population during which the beneficiairies are identified and consulted as to which elements from a menu of options should be included, when and where, and subsequent decisions taken with regard to details of programme implementation.

The programme approach is indicated by the umbrella character of the project organization, with "sister" projects in several countries, as well as a number of additional measures which are intended for the best possible exchange of information leading to the sharing of experiences and ideas for new approaches. On a regular basis, seminars and workshops are held, where project personnel meet and hold discussions, study trips are arranged to "sister" projects for other relevant projects) and strategies are formulated. The village/community based organizational structures, which the projects seek to form, if they are not yet in existence, are naturally an integrated part of this process.

The advantages of this approach are evident: not only do projects not necessarily have to repeat all the failures of earlier, similar projects, but the best from the past can be exploited in a greater context for the future development of designs and approaches. Furthermore, this approach offers donors and agencies the institutional memory which is so often missing or lost.

It is too early to tell whether these good intentions will bear fruit. But recently UNSO arranged a workshop in Burkina Faso on "Resource management and Public Participation", which was a first and, according to all accounts, successful venture of this kind.

UNSO's experience : a summary

UNSO has in the past constantly tried to modify and redirect project approaches and designs in accordance with experience gained. In certain cases, project quality has definitely improved as a consequence. In other cases, bottlenecks and shortcomings have been caused by factors outside UNSO's control and with only limited possibilities for UNSO to counteract.

This process of learning and adaptation is, of course, continuous and a prerequisite for spearheading activities and developing innovative approaches.

For a small organization the size of UNSO (and a mandate so huge and complex), it is imperative that much effort be put into sharpening the focus and optimizing the impact and possibilities of replication. Given the complex nature of the problems at hand in the Sudano-Sahelian region, this is a valid ambition but extremely difficult to put into practice. In more optimistic moments, UNSO might claim a few cases where we have spearheaded project designs/activities, which were subsequently adopted by other organizations/agencies as their own.

UNSO's experience from operational projects and programmes in the Sudano-Sahelian region is probably not much different from that of other agencies. The lessons drawn may be summarized as follows:

- although much more difficult to achieve, multisectoral and more integrated approaches are needed, to deal simultaneously with several aspects related to the complex and interwoven set of factors causing resource degradation, however maintaining a "project nucleus";

- smaller scale and technologically not too sophisticated projects are needed, which can be managed and have better scope for replication by the local population after they have been completed;

- indigenous technologies (and species) should be further utilized, researched and exploited, in cases combined with the introduction of adapted and suitable foreign technologies;

- while seeking to preserve/restore/rehabilitate natural resources by more appropriate management principles and other project interventions, project results should be of immediate material benefit to the local population;

- although it might complicate the management and running of projects and prolong project life, public participation in design and implementation of projects/programmes is imperative and should be given the highest priority;

- the implication of efficient participation by the local population might consist of operating with a "menu of options" project design strategy, where the introduction of elements from the "menu" in operational activities takes place only after appropriate consultations with the local population;

- several projects assembled under some form of programme approach might constitute a feasible alternative in fostering harmonization, co-ordination and sharing of experiences gained, as compared to the punctual (and often sectoral) prosect approach;

- the basic elements in natural resource management projects are agro-forestry measures and soil and water conservation techniques, combined with more appropriate rules and regulations pertaining to the local population's rights and its access to land and natural resources in general;

- Government policies are often adversely affecting efforts in introducing efficient natural resource management principles as the lack of incentives and material gain from project involvement might threaten the project#s viability and sustainability, and cause further ecological degradation;

- the objective of natural resource management projects is to protect the <u>productive</u> natural resource base and foster productivity increase in agriculture and livestock, in a sustainable and ecologically sound way. Only with productivity increases can the vicious circle of poverty, resulting from environmental degradation, and further accelerating the process of degradation, have a chance to be reversed.

Planning, co-ordination and monitoring

As mentioned above, Government policies are, in a number of cases, all determining when evaluating the outlook for the various kinds of interventions introduced and, ultimately, for the prospects of restoring some kind of ecological balance. This applies to pricing and marketing policies and policies related to the use of land and other natural resources. It also applies to problems related to internal conflicts and civil strife, as well as conflicts among nations.

The role of Government and Government institutions is crucial, also in a more positive way. Assistance to strengthening national capacities in planning, co-ordination and monitoring of the process of ecological degradation is a prerequisite for successful project and programme implementation, and is needed as a necessary follow-up and as a method of synthesizing experiences gained.

Following the adoption in 1977 of the world Plan of Action to Combat Desertification, UNSO has in a number of its mandate countries assisted in formulating National Plans of Action to Combat Desertification. While these Plans have been useful tools, particularly in providing a reference point for a set of concrete project activities ready for financing, it appears that there is an even greater need for technical and planning advice on a <u>continuous</u> basis, addressing in a more comprehensive way the <u>planning process</u> at large.

The objective of UNSO's assistance to Governments in formulating strategic frameworks and, in various ways, supporting the <u>process</u> of planning has been:

-   to view the process of planning in a more coherent and comprehensive way, emphasizing continuity, sustainability and long-term impact;

-   on the basis of the above, to provide assistance to planning and co-ordination activities by supporting various elements in the stages of the planning process, using a flexible and differentiated approach in accordance with country-specific factors and needs;

-   to support the Governments' capability for coordinating efforts in countering the process of desertification/ecological degradation, with the aim of economizing on scarce resources and avoiding duplication of efforts;

-   to indirectly assist the Governments' efforts in coordinating the fight against drought and desertification by promoting <u>international</u> co- ordination among donors and agencies/organizations, to economize on scarce resources, avoid duplication of efforts and improve on quality;

-   to advise Governnents on questions related to the monitoring of the process of ecological degradation and/or to provide inputs enabling Governments to better monitor the situation, thereby improving the possibility of implementing correct and realistic planning procedures;

-   to ensure that the assistance to Governments in planning, co-ordination and monitoring be conceived with a broader view in mind by forming part of UNSO's other project/programme activities in such a way that the two programme areas be mutually supportive and interactive.

Apart from preparing strategic frameworks, such as the National Plans of Action to Combat Desertification, UNS0 has particularly emphasized the strengthening of Government planning and monitoring capacities, by assisting in the establishment of co-ordinating, inter-ministerial committees or units, erecting smaller permanent planning or monitoring bodies, providing experts or consultants and various types of equipment such as computers and software packages (in particular GIS), establishing data bases and information systems, conducting training and education seminars, encouraging networking etc., etc.

In the field of monitoring, UNSO (with DANIDA financing) has supported the establishnent of the Ecological Monitoring Centre in Dakar, Senegal. The CSE uses relatively costly and sophisticated techniques, such as the analysis of satellite images (for the production of maps related to vegetation cover, watering holes, soil structures, etc.), low flying reconnaisance flights (enumerating livestock and livestock movements, validating information received and analysed by satellite images, etc.), ground truth observation (measurements in the field of vegetation types and biomass quality and other features, to validate the analysis carried out in previous steps), combined with socio-economic analysis of household herding and watering practices, problems of overgrazing, etc.

From monitoring a predominantly pastoral region (the Ferlo), the Centre has gradually developed its range of activities to also include major agricultural areas, and has the potential of becoming a truly national monitoring centre. Its role in the region with regard to training and teaching specialists from other countries should also be mentioned.

However, this approach to monitoring might only be justified in Senegal where the Centre has demonstrated the various technologies and applications possible and will be able to act as a centre of excellence within its field, serving not only Senegal but also other countries of the region, e.g. in training and research. The costs involved and the relatively high degree of sophistication of the techniques used, have convinced UNSO that in other countries one should rather encourage the establishment of smaller monitoring units, gradually building up the monitoring capacity in a country in accordance with needs expressed and possibilities of mastering the techniques used. This more longterm approach might better serve the countries in question, as training of personnel and the installation of equipment and hardware go hand-in-hand, in accordance with the needs expressed.

Note:
The value of UNSO's actual project portfolio is well over US $ 100 million and its annual turnover is approximately US $ 30 million. UNSO's main contributors (around 70 % of the total) are the Scandinavian countries (Sweden, Denmark, Norway Finland, in that order), followed by Holland, Canada, and others.

Furanocoumarins, potential molluscicides or just phototoxic compounds?

Leon Brimer

Abstract

In recent years, natural compounds with molluscicidal effect have been discovered at an accelerated pace. Most are never investigated in any detail concerning their kinetics and dynamics in humans and other higher non - target organisms. The furanocoumarins, also used in the medical treatment of psoriasis are among the few molluscicidal plant constituents for which some knowledge about absorption, distribution and excretion has been gained. However, the ability of certain of these compounds to cause a phototoxic response, and to cause mutations in conjunction with UVA - radiation has - on the other hand - limited the research within their molluscicidal effect. Consequently, it is interesting to note that the structure-activity relationship differs for the molluscicidal and for the phototoxic effect of these compounds respectively. This fact underlines the importance of further research within the molluscicidal effect of furanocoumarins and its biochemical mechanism.

Molluscicides, the current status

In the combat against human and domestic animal schistosomiasis (bilharziasis) and other parasitic diseases with water related snails as obligate intermediate hosts (e.g. infections with <u>Fasciola hepatica</u>, domestic animal liver flukes) molluscicides have played an ever changing yet important rôle (1-2). The result of years of laboratory screening, field testing, and evaluation of actual snail reduction campaigns is, that to - day only one compound, the synthetic **niclosamide** (2',5 - dichloro - 4'- nitrosalicylanilide, Bayluscide[R], Mollutox[R]) is in widespread use (2,3). This situation is totally different from that of most other classes of pesticides (4), and makes no room for individually adapted - chemically based - eradication strategies. Moreover, niclosamide is regarded as a relatively costly compound by most developing countries.

Facing these realities, expert commities advising WHO as well as other leading scientists from 1983 (5) to 1986 (6) and onwards recommended further research and development within the field of plant derived molluscicides, notably the saponins from <u>Phytolacca dodecandra</u> L'Herit - the "endod" plant (7). As an integrated part of these initiatives, UNDP/World Bank/WHO supported the preparation of a book including a number of reviews concerning known plant molluscicides and molluscicidal plants (8).

Among the listed plant compounds (8) showing promising toxicity towards water related snails, none has been fully characterized concerning their toxicity towards higher animals and humans. This is equally true for the ecotoxicological aspects if used in open field systems i.e. applied to or brought into the environment. Thus concerning the Phytolacca saponins, the "WHO Endod toxicology expert group" had to recommend, that "additional studies should be performed to confirm the data already on file, and to provide the essential data for the basic regulatory requirements internationally adopted for the first level (Tier 1) at the present time" (6). However, two of the compounds, i.e. 9 - methoxy - 7H - furo (3,2g)(1) benzopyran - 7 one (= xanthotoxin, ammoidin, methoxalen, oxsoralen, 8 - methoxypsoralen, 8 - MOP) and 4 - methoxy - furo - (3,2g) chromen - 7 - one (= bergapten, 5 - methoxypsoralen, 5 - MOP), both belonging to the group of linear furanocoumarins (furocoumarins, psoralens) fig. 1, have been subjected to at least certain investigations both as plant toxins - which they are - and as drugs for the treatment of e.g. vitiligo and psoriasis, for a score of years and more.

Linear furanocoumarins and their biological effects

The compound 8 - MOP has been more or less routinely used in the treatment of vitiligo world-wide since 1953, and of psoriasis - both in the United States of America and in Europe - since 1974 (9), while 5 -MOP has been investigated as a potential drug for the treatment of psoriasis (10). Consequently, the absorption, distribution, metabolism and excretion of these closely related compounds have been studied (11-12), as have their pharmacological and toxicological mechanisms and effects. Being the only compounds within the listed plant constituents for which this is true, it might seem natural to summarize that linear furanocoumarins constitutes an ideal group of compounds for further research and development as mulluscicides. However, the conclusion in the aforementioned book (8) was "use of the coumarins and furanocoumarins as molluscicides is problematic because of their pharmacological and toxicological effects - especially their hepatotoxic and photosensitizing effects - which have been extensively documented" (13, Henderson et al.).

The phototoxic (photosensitizing) activity mentioned by Henderson et al. (13) is a characteristic effect of several of the known linear furanocoumarins, both natural and synthetic, when given (orally or topically) to higher animals together with ultraviolet (UV) irradiation. This skin response, which macroscopically may include erythema, blistering, hyper pigmentation and thickening of the stratum corneum (14, 15) is the basis for the often used designation of this group of compounds as "phototoxic furanocoumarins". Phototoxicity has also been described for this group of compounds towards bacteria, fungi and protozoans (these organisms are killed or their growth is inhibited). The above mentioned hyper pigmentation which occurs both on normal skin and on vitiligo suffering skin, is the basis for the very early use (16) of these compounds together with sunlight in the treatment of vitiligo. The more recent incorporation of e.g. 5 - MOP containing citrus oils (notably bergamot oil from Citrus bergamia Risso et Poiteau) in various commercial sunscreens (17,18) is due to the same phenomenon. The mechanism behind the clearing of psoriatic lesions as observed in combination with UVA (wavelength 320 - 400 nm) light, has long been hypothesized to be due to the ability of these compounds to form DNA adducts following this UVA irradiation (19). By clearing is meant a suppression of the abnormally high growth rate of the keratinocytes.However, although inhibition of DNA synthesis have been observed both "in vitro" (20) and "in vivo" (21) following this regime, DNA - psoralene adducts (mono - and di(bi)adducts (22)) have only been characterized "in vitro". Thus adducts was formed as a result of a reaction between isolated DNA (23) or synthetic DNA-fragments (24) and a linear (or angular) furanocoumarin, following UVA radiation, i.e. these adducts have never been isolated from PUVA - treated human skin (19). In agreement with this fact other mechanisms have also been proposed for the keratinocyte growth suppression (19).

In addition to the phototoxicity, Henderson et al. (13) also mentioned the hepatotoxic effect of the furanocoumarins. It is certainly true, that both coumarin (1,2 - benzopyrone, cis - o - coumarinic acid lactone) and some of its derivatives such as certain linear furanocoumarins, have been reported to cause hepatotoxicity (25-27). Thus coumarin was given at a concentration of 0.5% w/w in the diet of male Sprague - Dawley rats for 18 month, which resulted in excessive development of cholangiofibrosis (25), while 8 - MOP was administered orally in corn oil to male and female F 344/N rats at doses from 25 - 400 mg/kg/day for 13 weeks (26). The latter experiment showed effect on body weight at 100 mg/kg/day, and affection of liver and testes from 200 mg/kg/day. This 8 - MOP investigation, performed after the conclusions reached by Henderson et al. (13), was done as part of the USA "National Toxicology Program" - having as one of its guidelines "the testing of old chemicals for which systematic testing has not been performed". The early and sparse reports on liver damage after PUVA treatments (28), have probably been part of the basis for the conclusions of Henderson et al..

Linear furanocoumarins as potential molluscicides

As natural compounds with molluscicidal activity (29-31) a few of the linear furanocoumarins, notably 8- MOP and 5 - MOP (29) and imperatorin (ammidin) (30, fig. 1.) have been rated for their molluscicidal activity. Unfortunately no clear conclusion can be drawn from the literature, due to differences in the snail species, sizes, duration of exposure and period of reconstitution. More important, none of the authors mention the lighting conditions during the experiments. However, if one compares the published results, firstly, to one another, and secondly to the relative acute phototoxic activities, as reported for the same compounds, an interesting picture emmerges. Table 1 shows that 8 - MOP and 5- MOP are nearly equally potent in their phototoxic properties, while the activity of imperatorin - as judged from the relatively few investigations made on this compound - seems less potent. If one restricts the discussion to the phototoxic skin reaction on higher animals and humans (investigations I - VI), it further becomes clear, that 8 - MOP is more potent in this respect than 5 - MOP. A comparison of these data with those for the reported molluscicidal activity, gives the rankings shown in table 2.

Thus it appears, that the quantitative structure activity relationship concerning the phototoxic skin response on humans (whether the compounds be taken orally or given subcutaneously or administered topically) does not correlate with that reported for the molluscicidal activity towards water related snails.

Figure 1:

Structural formulae for some linear furanocoumarins (psoralens). A: I **psoralen**. B: compounds that have been shown to be molluscicidal (references given in brackets); IIa R_1 = H / R_2 = OCH_3, bergapten (5-MOP, 29); IIb R_1 = OCH_3 / R_2 = H, xanthotoxin (8-MOP, 29); IIc R_1 = R_2 = OCH_3, isopimpinellin (29); III imperatorin (30); IV chalepensin (13,31). C: compounds that are used in PUVA treatments (in addition to the compounds IIa,b); V trioxalen (trimethylpsoralen (TMP), 36,37); VI 3-cabethoxypsoralen (3-Cps, 45).

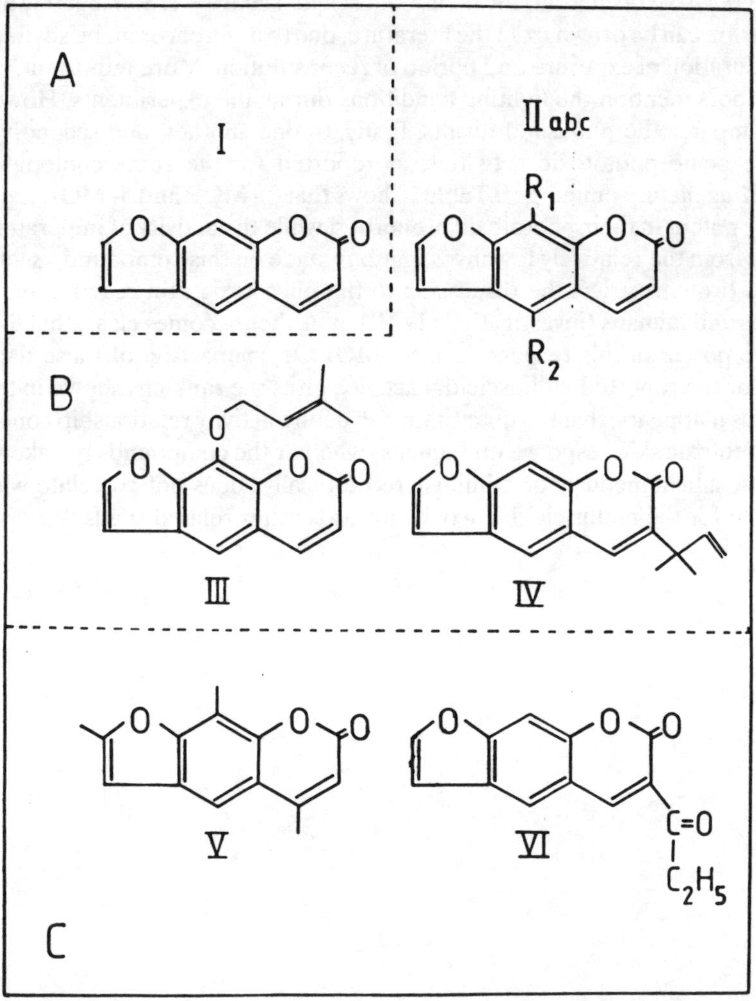

Table I:

The relative phototoxicity of selected linear coumarins as reported in the literature.

a. For the higher animals based on the formation of erythema, when nothin else is indicated.

b. References to the investigations: I (18), II (22), III (48), IV (20), V (25), VI (49), VII (50), VIII (51), IX (52).

c. Abbreviations for the test organisms: c = chicken; h = man; g = guinea pig; m = mouse; p/f: p = protozoans (Paramecium caudatum and Tetrahymena pyriformis)/ f = fungus (Candida albicans); f = fungi (Candida spp.); f/b: f = fungus (1 Candida albicans, 3 Saccharomyces cerevisiae)/ b = bacteria (2 Micrococcus lutens, 4 Escherichia coli).

d. Inhibition of epidermal DNA synthesis in vivo. Rating of phototoxicity: 1 = strongest; 4 = weakest, but positive; (-) = negative (under detection limit); n = not determined.

TABLE I							INVESTIGATION [a,b]					
						ORGANISM						
COMPOUND	I	II	III	IV	V	VI	VII	VIII	IX			
	h	h	h	c	m	g	P/f	f	f/b			
									1	2	3	4
XANTHOTOXIN (8-MOP)	1	1	1	1	1	2	2	2	2	3	2	2
BERGAPTEN (5-MOP)	2	2	2	2	2	h	1	1	3	2	3	2
IMPERATORIN	n	n	-	3	n	n	n	n	n	n	n	n
ISOPIMPINELLIN	n	n	-	4	n	n	n	n	-	-	-	-
CHALEPENSIN	n	n	n	n	n	n	n	n	n	n	n	n
4,5',8-TMP	n	n	n	n	n	1	n	n	1	1	1	1

TABLE II	EFFECT	
COMPOUND		
	PHOTOTOXIC	MOLLUSCICIDAL
8-MOP	1	3
5-MOP	2	1
IMPERATORIN	3	2

Table II:
   The relative phototoxicity of selected furanocoumarins (Table I) compared to the relative molluscicidal effects as judged from the literature data (1 = most toxic).

Molluscicidal contra phototoxic effect

The fact that some linear furanocoumarins seems to possess a reasonable molluscicidal activity, together with:

1. The lack of knowledge concerning the correlation - if any - of this effect, with that of the well known phototoxic properties of some of these compounds.

2. The missing reports on phototoxic effects on target snails.

3. the (after all) relatively moderate <u>general</u> toxicity, as compared to e.g. most organophosphororus and carbamate   pesticides in current mass scale use (32), seen in a 13 weeks experiment (26) for the even strongly photoactive 8 - MOP.

- prompted me to investigate the molluscicidal activity of the three linear furanocoumarins 8 - MOP, 5 - MOP and imperatorin, as seen in the presence or absence of UVA light in more detail.

Materials and methods

a: snail material, chemicals/water quality, and instruments.

1. (Snails): The target snail used in these investigations were Biomphalaria glabrata Say (albino strain from Puerto Rico, laboratory breed of the Danish Bilharziasis Laboratory, originally received through the London School of Tropical Medicine and Hygiene).

2. (Chemicals and water quality): 1,2 - Propanediol (propylenglycol) were USP quality. The three furanocoumarins were obtained from Carl Roth (C. R.), Karlsruhe, FRG: bergapten (5-MOP), catalog no. 2-7006; imperatorin (ammidin/marmelosin), no. 2-5816; xanthotoxin (8-MOP), no. 2-5497. All experiments were performed using "standard snail water" with a Ca^{+} $^{'}$ conc. of 0.1 mM (33).

3. (Instruments): UVA - irradiation was obtained from a Camag (Muttens, Switzerland) model 02.29230 thin-layer chromatography (TLC) lamp, which is equipped with one 8 W low - pressure mercury tube and a filter against visible light. The lamp was used at nominal wavelength 366 nm. The actual spectrum and the total integrated energy for this light source were measured (fig. 2, 34).

b: methods

1. (General): The general dark experiments and all snail recovery were done in 200 ml polystyrene beakers with perforated covers, while experiments involving UVA - irradiation, as well as experiments to be directly compared to these, were run in a special set-up (ref. below). The recovery period was 72 hours in all experiments, as calculated from the termination of the last treatment, whether chemical or physical, of the test animals. All recoveries were run in the dark,as described under "determination of molluscicidal activity".

Figure 2:
Measured spectrum of the light source used in this study. Measurements were
done 84 mm from the lamp window. Integrated (total) energy at this distance
1025 micro W/cm₂.

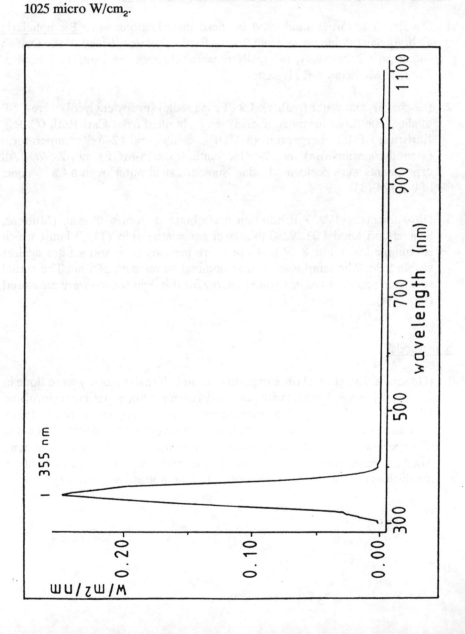

2. (UVA - set-up): The UVA - irradiation set-up consisted of a rectangular glass container 170 x 220 mm (height 60 mm), in which a thin nylon net was distended - by means of a frame - 25 mm above the bottom. The net quality was Nytrel - TY 250 HD^R (polyamide 6.6, open space 37%, L'Union Gazes A Bluter S. A. , Panissieres, France). One litre of the test solution was used, resulting in a water depth of 26 mm (one mm above net level). The container was placed under the light source, the distance from lamp window to water surface being 85 mm. The test snails, kept under the water surface by the net, were restricted in their horizontal movement to a central area of 110 x 170 mm by means of a rectangular polystyrene frame, from bottom to net top. The frame was perforated with 80 holes 4mm in diameter, to allow free water circulation in the bath. This set-up resulted in a very uniform UV intensity over the water surface.

3. (Determination of molluscicidal activity - in the dark): The three compounds: bergapten, imperatorin, and xanthotoxin were all evaluated for their molluscicidal activity as described by WHO (35). Thus 5 snails (sized 6 - 8 mmm) were exposed to a given concentration of the compound in question, in a container holding 200 ml of standard snail water for a period of 24 hours. This was done 4 times for each concentration (a total of 20 snails). The compounds were all dissolved with the help of 1% v/v of 1,2 - propanediol. Three times twenty snails subjected to 1% v/v 1,2 - propanediol for the same period resulted in a kill rate of 0%. In all experiments the containers were wrapped in aluminium foil, to protect from light, both during the exposure and during the recovery period. The temperature in the beakers was throughout the experiment measured to be 21 - 22 °C.

4. (Potentiation of molluscicidal activity by UVA - light):

The effect of combined action of molluscicidal furanocoumarins and UVA - irradiation, was studied.

In a preliminary experiment 5 snails in 200 ml of snail water containing 1% v/v of 1,2 propanediol, were subjected to one of the following treatments, duration 2 hours: a, UVA - irradiation; b, 10 ppm of 8 - MOP; c, UVA - irradiation + 10 ppm of 8 - MOP (simultaneously). The three experiments were performed twice (series 1 and 2).

In a subsequent study, the following experiments were included:

a. 8 - MOP: 1o snails were exposed to 15 ppm of 8 - MOP (1,2 - propanediol, 1% v/v) for 2 hours (in the dark), and subsequently, after a period (t) of rest in fresh snail water (40 ml per snail, dark conditions), to two hours of UVA - irradiation also in fresh snail water.

The periods of resting were: t = 0 - 2 - 4 and 16 hours. The results of these experiments were compared to those of an experiment with irradiation of the snails when exposed to the 8 - MOP (as in the preliminary trials).

b. Imperatorin: 10 snails were subjected to 12 ppm of imperatorin + UVA - irradiation (simultaneously) for 2 hours, this experiment was done in dublicate.

Results and discussion

Experiments with controlled light conditions are clearly providing an insight into the molluscicidal properties of the furanocoumarins, that is not possible to attain through other studies. Thus the molluscicidal activities reported so far in the literature (Table 2, and references 29 and 30) reflect mainly the dark toxicity as seen by comparison with the measurements shown in Table 3. Further, a phototoxic response (as reported for many other organisms) also exists for snails when exposed to certain linear furanocoumarins in UVA - irradiated water as proved by the results presented in Fig.3. A comparison between Table 1 and the experimental results in Table 4 indicates that the structure activity relationship concerning the "photomolluscicidal" effect - in this study only reflected through an endpoint assay - might be similar to that seen for other organisms. The mechanism underlying these phototoxic responses has been intensively studied and discussed for many of the organisms presented in Table 1, as already stressed. However, the mechanisms underlying the dark toxicity towards water related snails have not been investigated.The present theoretical as well as experimental study supports the hypothesis that the structure activity relationship for this latter toxicity might be different from that of the photoresponse; conclusion based on Tables 2 to 4. Finally it should be noted, that the experiments in this study do not indicate that the phototoxic furanocoumarins (8 - MOP taken as an example) are taken up by and distributed within the snail , or that the compound adher to the skin , to an extent or in a way that facilitates a phototoxic response when irradiated after removal from the compound (Table 4).

Table III:
Molluscicidal activity of linear furanocoumarins in the dark. Twenty snails exposed in 24 h; mortality rated after 72 h of recovery in pure snail water. The toxicity of imperatorin as measured over 24 h being limited by the solubility, i.e. a saturated solution was obtained at a concentration around 18-20 ppm.

TABLE III COMPOUND	% MORTALITY AT DIFFERENT CONCENTRATIONS IN ppm ($mg \cdot l^{-1}$)																
	1	2	3	4	5	6	8	9	10	12	15	18	20	24	30	40	60
5-MOP	0	5	10	70		100	100		100								
IMPERAT.			0				0		20		40	40	55		55		
8-MOP				0					0				10			60	90	100

Table IV:
The combined effect of the furanocoumarins (8-MOP and imperatorin respectively) and UVA-irradiation. The effect of a time gab between the exposure to xanthotoxin and to the irradiation. Xanthotoxin: number of snails killed out of 10 (single determinations). Imperatorin: number of snails killed out of 10 (in dublicate).
Treatment regime: (sim. = simultanious) means that the snails were irradiated in the solution containing the furanocoumarin; (0) means that the snails were exposed to xanthotoxin in the dark, and immediately after this to the UVA light, when situated in pure snail-water.
Exposures: Chemical 2 h; UVA light 2 h.
Recovery: 72 h, observe that 12 ppm of imperatorin cause a higher mortality over 24 h of exposure (in the dark) that do the 15 ppm of 8-MOP.

TABLE IV COMPOUND	"TREATMENT": HOURS BETWEEN CHEMICAL EXPOSURE AND UVA-IRRADIATION				
	sim.	0	2	4	16
8-MOP (15 ppm)	7	0	0	0	0
IMPERATORIN	0/1	n	n	n	n

Figure 3:
Molluscicidal effect of 10 ppm xanthotoxin (8-MOP) and/or UV light irradiation (1025 micro W/cm₂). Two hours of treatment, 72 hours of recovery; 5 snails in each series. Each experiment in duplicate (5 snails in each experiment).

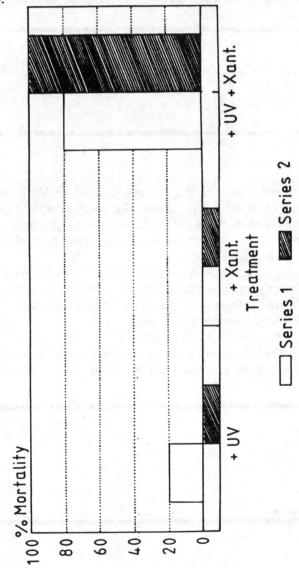

The present study contributes to the discussion of whether coumarins (and notably furanocoumarins) constitute a structural group of potential value in the search for molluscicides. However, the mutagenic and carcinogenic properties described for certain of these compounds, certainly represents a drawback from the very beginning.

The ability to form adducts with DNA is most probably the explanation for the photomutagenic action demonstrated by some of these compounds. Thus both 8 - MOP and the synthetic compound 2,5,9 - trimethyl - 7 H - furo(3,2g) chromen - 7 - one (4,5',8 - trimethylpsoralen, 4,5',8 - TMP, Trioxalen, Trioxysalen, fig. 1.) which is also used in PUVA treatments (36) is mutagenic in conjunction with UVA light in several microbial test systems (37). Furthermore both compounds proved to be mutagenic in the dark in an E. coli system , after activation by rat liver microsomes (37).

In agreement with the above mentioned mutagenic potential, 8 - MOP have been found to be carcinogenic in conjunction with UVA light under specified conditions, whether given orally, interperitoneally or administered topically (38).

This carcinogenic potential may be the explanation for the reported triple enhancement (over 5 years) of cutaneous carcinoma in patients subjected to PUVA treatments in the USA, as reported by Stern (39). This report ,though not the first (40), was among those that prompted scientists as well as administrators to discuss the possible risks related to: a solarinduced cancer, b cancer originating from PUVA treatments of psoriasis and c an enhancement of cutaneous cancer due to the use of suntan preparations based on 5 - MOP (41).

Due to this photomutagenic/photocarcinogenic activity 5 -MOP was voluntarily withdrawn from commercial sunscreens in the U.K. (42).

Carcinogenic compounds of any kind are clearly unacceptable as molluscicides. Nevertheles it seems interesting to note, that the reported higher frequency of certain cancers in PUVA treated Americans (39) seems to originate in the treatment regime as such, since examinations in Europe did not show the same enhancement (43). Furthermore certain experts in toxicology have recently gone so far as to consider a renaissance for 5 -MOP in sunscreens for certain skin types - after a balanced evaluation of risks and benefits (17). Several compounds are useful as screens against UVB radiation (290 - 320 nm), while only few well tested agents are photoprotective against UVA (44). It is further interesting to note, that while most scientists originally ascribed the quality of clearing psoriatic lesions in PUVA treatments only to furano-coumarins that could cause the previously described phototoxic response (identical to those which can form biadducts with DNA), subsequent screenings have shown a different picture. Thus, 3 - carbethoxypsoralen (a linear furanocoumarin (fig. 1) that can form only monoadducts with DNA) was shown to be effective in the clearing process too, although neither erythema nor hyper pigmentation were induced (45). Although still mutagenic in certain tests, this as well as other so-called monofunctional furanocoumarins tested later seems to be less- or maybe non-carcinogenic (45). Among the tested monofunctional compounds able to clear psoriatic lesions in conjunction with UVA - irradiation, one will even find angular compounds such as 4,5' - dimethylangelicin (DmAgn) (46) and 6,4,4' - trimethylangelicin (TMA) (23). Altogether a picture that stimulates further research also in the field of molluscicidal furanocoumarins.

To summarise, the data discussed seem to indicate that coumarins/ furanocoumarins deserve a closer investigation as potential pesticides in general, a trend already in progress (47), and as molluscicides in particular. It might be possible to separate the molluscicidal and the phototoxic/carcinogenic properties of these molecules in the future.

References and notes

1   O. Tameim, Z.B. Zakaria, H. Hussein, A.A. Gaddal, W.R. Jobin, J. Trop. Med. Hyg. **88**, 115 (1985).

2   G. Webbe, In Plant Molluscicides ,K.E. Mott Ed. (John Wiley & Sons LTD, Chichester ,1987), pp. 1 - 26.

3   V.K. Brinkmann, C. Werler, M. Traore, R. Korte, Trop. Med. Parasitol. **39**, 157 (1988).

4   C.R. Worthing, S.B. Walker (Eds.), The Pesticide Manual, A World Compendium, 8'th Ed. (British Crop Protection, Thornton Heath (1987) pp. 1 - 851.

5   D. Heyneman, In Phytolacca dodecandra (Endod), A. Lemma, D. Heyneman, S.M. Silangwa ,Eds. (Tycooly Int. Publ. Ltd., Dublin 1984), pp. ix - xiii.

6   Anon., In Endod II (Phytolacca dodecandra), L. Makhubu, A. Lemma, D. Heyneman, Eds. (Council of Int. Publ. Aff., New York 1987), pp.131 - 148.

7   H. Kloos, In Studies on the molluscicidal and other properties of the Endod, A. Lemma, D. Heyneman, H. Kloos, Eds. (Department of the Navy, Arlington 1979), pp. 20 - 61.

8   K.E. Mott (Ed.), Plant Molluscicides (John Wiley & Sons, Chichester 1987).

9   J.A. Parrish, T.B. Fitzpatrick, L. Tanenbaum, M.A. Pathak, N. Engl. J. Med. **291**, 1207 (1974).

10 H. Hönigsmann, E. Jaschke, F. Gschnait, W. Brenner, P. Fritsch, K. Wolff, Br. J. Dermatol. **101**, 369 (1979).

11 M.A. Pathak, T.B. Fitzpatrick, J.A. Parrish, R. Biswas, In Proc. 7'th Int. Congr. on Photobiology, A. Castellani, Ed. (Plenum Press, New York, 1977), pp. 267 - 281.

12 S.J. Kolis, T.H. Williams, E.J. Postma, G.J. Sasso, P.N. Confalone, M.A. Schwartz, Drug Metab. Dispos. 7, 220

13 T.O. Henderson, N.R. Farnsworth, T.C. Meyers, In Plant Molluscicides K.E. Mott, Ed. (John Wiley & Sons, Chichester, 1987), pp. 111 -

14 R. Brickl, J. Schmid, F.W. Koss, Photodermatol. 1, 174 (1984).

15 E.A. Emmett, in Casarett and Doull's Toxicology, the Basic science of Poisons (Macmillan Publ. Comp., N.Y.,1986), pp.412 - 431.

16 G.W. Ivie, J. Agric. Food Chem. 26, 1394 (1978).

17 A.R. Young, C.S. Potten, C.A. Chadwick, G.M. Murphy, A.J. Cohen, Pigment Cell Res. 1, 350 (1988).

18 S.T. Zaynoun, B.E. Johnson, W. Frain-Bell, Br. J. Dermatol. 96, 475 (1977).

19 E. Lee, Dissertation (Rutgers The Stare University, New brunswick, 1987; Avail. through UMI Dissertation Inf. Service) pp. 1 - 222.

20 J.F. Walter, J.J. Voorhees, W.H. Kelsey, E.A. Duell, A. Arbor, Arch. Dermatol. 107, 861 (1973).

21 F. Bacchichetti, C. Antonello, F. Carlassare, O. Gia, M. Palumbo, F. Bordin, Photodermatol. 2, 221 (1985).

22 G.D. Cimino, H.B. Gamper, S.T. Isaacs, J.E. Hearst, Annu. Rev. Biochem. 54, 1151 (1985).

23 G.Miolo, F. Dall'Acqua, E. Moustacchi, E. Sage, Photochem. Photobiol. 50, 75 (1989).

24 A.T. Yeung, B.K. Jones, C.T. Chu, Biochemistry 27, 3204 (1988).

25 J.G. Evans, E.C. Appleby, B.G. Lake, D. M. Conning, Toxicology 55, 207 (1989).

26 J.K. Dunnick, M.R. Elwell, Toxicology 56, 123 (1989).

27 R. Kramer, E. Kaiser, Toxicon 6, 145 (1968).

28 M. Bjellerup, M. Bruce, A. Hansson, G. Krook, B. Ljunggren, Acta Derm. Venereol. (Stockh.) 59, 371 (1979).

29 A. Schonberg, N. Latif, J. Am. Chem. Soc. 76, 6208 (1954).

30 S.K. Adesina, C.O. Adewunmi, Fitoterapia 56, 289 (1985).

31 C.O. Adewunmi, F. Delle Monache, Fitoterapia 60, 79 (1989).

32 G.J. Smith, Pesticide use and toxicology in relation to wildlife: Organophosphorus and Carbamate compounds (United States Department of the Interior, Resource Publication no. 170, Washington D.C., 1987), pp. 11-161.

33 H. Madsen, J. Appl. Ecol. 24, 823 (1987).

34 Light spectrum and intensity were measured by means of a portable spectroradiometer model LI - 1800, LI - COR, Lincoln, Nebraska, U.S.A.. Measurements were done at the normal water level in the actual set - up, i.e. 84 mm from the lamp window.

35 Anon., Bull. WHO 33, 567 (1965).

36 J.E.F. Reynolds (Ed.), Martindale, the Extra Pharmacopoeia, 28'th ed. (The Pharmaceutical Press, London, 1982), pp. 1657.

37 D.J. Kirkland, K.L. Creed, P. Mannisto, Mutat. Res. 116, 73(1983).

38 B.A. Bridges, M. Greaves, P.E. Polani, I. Wald, Mutat. Res. 86, 279 (1981).

39 R.S. Stern, N. Laird, J. Melski, N. Engl. J. Med. 310, 1156 (1984).

40 J.H. Epstein, N. Engl. J. Med. 300, 852 (1979).

41 W.M. Wardell, in Current concepts in cutaneous toxicity, V.A. Drill, P. Lazar, Eds. (Academic Press, N.Y., 1980) pp. 271 - 288.

42 A.J. Cohen, Food Chem. Toxicol. 27, 69 (1989).

43 N.K. Gibbs, H. Hönigsmann, A.R. Young, Lancet i, 150 (1986).

44 M.A. Pathak, J. Dermatol. Surg. Oncol. 13, 739 (1987).

45 L. Dubertret, D. Averbeck, F. Zajdela, E. Bisagni, E. Moustacchi, R. Touraine, R. Latarjet, Br. J. Dermatol. 101, 379 (1978).

46 W. Gast, M. Rytter, C. Hofmann, U.-F. Haustein, J. Barth, T. Walter, Photodermatol. 2, 392 (1985).

47 B.B. Gupta, R.V. Singh, O.P. Malik, H.R. Kataria, Pestic. Sci. 21, 51 (1987).

48 L. Musajo, G. Rodighiero, Experientia 18, 153 (1962).

49 M.A. Pathak, J.H. Fellman, K.D. Kaufman, J. Dermatol. 35, 165 (1960).

50 A.R. Young, J. Barth, Photochem. Photobiol. 35, 83 (1982).

51 E.A. Knudsen, Photodermatol. 2, 80 (1985).

52 M.J. Ashwood-Smith, G.A. Poulton, O. Ceska, M. Liu, E. Furniss, Photochem. Photobiol. 38, 113 (1983).

53 I wish to thank the Director of the Danish Bilharziasis Laboratory, Dr. N. O. Christensen, for permission to work at the laboratory although I am officially employed at The Royal Veterinary and Agricultural University. I kindly thank Ass. Prof. J. Lemmich, Dep. Org. Chem., Royal Danish School of Pharmacy, for samples of imperatorin for pilot studies, Prof. Folke Rasmussen (Dep. Pharmacol. & Toxicol., The Royal Vet. - & Agricul. Univ.) and M. Sc. K. B. Ginn (Bruel & Kjaer) for valuable discussions concerning the manuscript and Ass. Prof. J. C. T. Petersen from the Dep. of Soil and Water and Plant Nutrition (The Royal Vet. - & Agricul. Univ.) for doing the light measurements.

2. Abstracts and short communications of ongoing activities

Non-poisonous methods of pest control during forest seed collection, handling, storage and transfer.

Henry Mourier

The problem

Seed-eating insects cause considerable losses to seeds of various leguminose tree species of importance in reforestation programmes in arid and semiarid zones of the tropical world. By far the most important pests are Bruchidae (seed beetles). The members of this family are for the most part associated with seeds belonging to the Leguminosae. Several species are important pests of beans and other pulses in storage, and in recent years there has been an increasing awareness of the infestation of seeds of forest tree seeds by beetles of this family.

In most instances the female beetle selects a ripening pod and lays an egg on the surface. The small larvae on hatching bores through the pod wall and seeks out a developing seed. It burrows into the seed and hollows out a chamber which it enlarges as it grows. The fullgrown larvae pupatea and later leaves the seed as an adult beetle, through a circular hole. Such parasitized seeds generally do not germinate. There is still lack of knowledge on taxonomy, distribution and biology of most bruchid species, and the knowledge on bruchid/ host relationship is also scarce.

More than thirty defence strategies - by the leguminouse plants that may be functional in eliminating or lowering bruchid destruction of the seeds have been listed, and as many adaptions evolved by the seed beetles, to circumvent these defences, are known. The effect on germination is obviously considerable, infestation rates of more than 90% are common, and the population dynamics of trees in the savanna ecosystem is no doubt influenced by these seed predators.

The quality of seed lots can be severely affected by the seed predating bruchid beetles, and control of these insects is essential to ensure a high percentage of viable seeds for planting programmes. In addition to this, the presence of living insects in seed in international transfer is not acceptable, and a matter of great concern to the quarentine authorities.

Various methods to solve this problem are in use, the application of heavy dosages of insecticides being the most common.

However, insecticide treatment of seeds harbouring insect larvae is not an effective control method, and apart from this the pesticides may be dangerous to human health and perhaps cause damage to the seeds.

The purpose of our study was to achieve information on simple alternative methods of control, and to determine their effectiveness as regards controlling pests of forest seeds, without damaging the seeds.

The work was carried out as a cooperation between the Centre National de Semences Forestières (CNSF) Burkina Faso, Danida Forest Seed Centre (DFSC) snd the Danish Pest Infestation Laboratory (DPIL).

Methods

Four different control methods were chosen for our preliminary investigations.

- Modified atmospheres. We observed the effect of flushing the plastic bags, containing the seeds, with CO_2, by a simple manual technique.

- Cold treatment. We investigated the effect of keeping the seed samples in an ordinary deep-freezer, at -18°C, for different periods of time.

- Heat treatment by microwaves. We determined the effectiveness of a commercially available microwave oven in controlling insects within seeds.

- Vacuum treatment. We observed the effect of keeping the samples at pressures between 20 and 50 mbar for 2,4,24 and 48 hours respectively.

Germination tests were performed at the DFSC.

Results

Flushing the plastic bags containing the seeds with CO_2 gave a >90% reduction with exposure time of one week. The germination tests confirmed that CO_2 treatment has no harmful influence on the germination power of the seeds. Exposure periods from one to six days at -18°C all gave more than 90% reduction in emergence, and the germination power was not harmfully influenced by the treatment. 3o sec exposure in the microwave oven only gave a 80% reduction in emergence. The germination tests showed declining germination with increasing dosage and the use of microwaves as a practical means of insect control, as regards seeds for sowing, was discarded.

As regards vacuum treatment, the percent reduction in emergence increased from 10 to 47, when exposure time was prolonged from 2 to 48 hours. This treatment does not have a measurable effect on the germination power of the seeds. No doubt vacuum treatment can eliminate infestation of seeds by bruchid beetles, provided pressure is sufficiently low and exposure period long enough, but exposure periods on that scale are not considered realistic for practical purposes.

Conclusions

The results indicate that <u>cold treatment and CO_2 treatment are possible alternatives to synthetic insecticides</u> for control of insect pests of forest seeds, but that further quantitative tests are required to establish data which can be translated into practical treatment scedules.

Management of forest resources in Senegal

Bienvenu Sambou

Abstract

During the last two decades, the accelerating degradation of forests in Senegal has lead to great concern in the local rural populations and administrative authorities. The on-going exploitation of wood, agricultural activities, bush fires, overgrazing, the droughts and forestry policy are the main reasons responsible for the degradation. The implementation of forest policy and other legislative measures has caused severe conflicts between authorities and local populations, and this has certainly not favoured the conservation of forest resources in Senegal.

Ecological effects of fuelwood plantations in Ethiopia.

Ib Friis and Anders Michelsen

Are *Eucalyptus* plantations in Ethiopia ecologicaly harmful, as it has been claimed
for similar plantations in other parts of the world? Current collaborational
research involving the Biology Department, Addis Abeba University, and the
University of Copenhagen looks for an answer to this question.

Eucalyptus globulus and other *Eucalyptus* species have widely been planted in
Ethiopia for the last 100 years in order to meet the strong demand for fuelwood
and construction poles. However, it has been claimed that *Eycalyptus* species
suppress the ground vegetation and affect the soil properties adversely. This in
turn should increase the risk of soil erosion and decrease long term productivity
of plantations, and thus be a non-sustainable land use. Several tree species other
than *Eucalyptus* have been established in plantations in Ethiopia, both exotic and
indigenous, and some of these may be a better choice from an ecological point of
view.

In the present research project 100 stands of various species of plantation trees
are investigated in about 10 plantation sites in western, southern, central and
eastern Ethiopia, as depicted on the map.

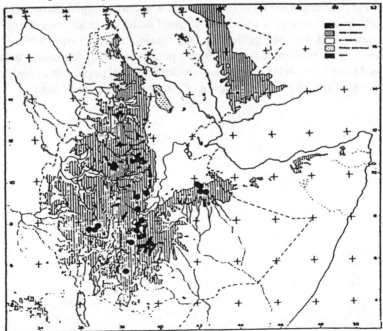

The investigated parameters include the species composition, coverage and biomass of the tree and the herb layers, light intensity, top and sub soil properties, fine root biomass, mycorrhizal colonization and root nodulation. Moreover the nutrient cycling is followed more intensively in 4 selected stands of *Eucalyptus globulus, Cupressus lusitanica, Juniperus procera* and a natural forest. The monthly leaf litter fall and litter decomposition rate in the forest floor is followed, together with the variation in ground vegetation biomass.

The collected data should provide material for ecologically based sound species selection and management of Ethiopian fuelwood plantations.

Conservation and development of the Air mountains and the Tenere Desert, Niger.

Kim Carstensen

In January of 1988 the Aïr-Ténéré area was declared a National Nature Reserve by the Government of Niger. The area comprises some 77,000 km², split into two relatively distinct environments, the Aïr Mountains and the hyper-arid Ténéré Desert.

The area is the home of about 4,500 people, of which some 2,500 are sedentary while the rest are nomadic or semi-nomadic. The population fluctuates, however, in relation to the availability of grazing resources.

Because of the area's geographical position between Sahara and Sahel, it is very diverse in a biological sense, compared to its more homogenous surroundings. In spite of serious over-utilization of the natural resources (mainly trees) during the droughts of the 70s and the 80s, the area has retained most of its biological values, and today it must be said to be quite unique in the region.

During the 3 years from 1987 to 1989, rains in the area were reasonably good, and trees and perennial grasses have clearly been recovering. At the same time both domestic stock and wildlife populations have been increasing.

Together with the IUCN - the World Conservation Union, WWF has been working in the Aïr-Ténéré area since 1979. Originally the project was a traditional animal protection activity (addax, ostriches, barbary sheep, gazelles, etc.), but gradually our approach has been broadened and our working methods have changed.

Maintaining the biological diversity of the area is still the main objective of the project, but we recognize that we cannot obtain this without cooperation with the local population. The people will have to be the guardians of the area in the long run, and therefore they have to be actively involved in the plans and activities promoted by the project.

The project focuses on the interface between the physical (soils, water, flora, fauna) and human environments, with a view to maintaining the biological diversity through the promotion of sustainable utilisation of the resources and the satisfaction of the needs of the human population.

In 1990 the project has entered a new phase with funding from Danida (through WWF-Denmark) and the Swiss Aid Agency, DDA (through IUCN). The phase will run until the end of 1993, with the main objective of establishing a process of participative planning leading to a management plan for the Aïr-Ténéré area:

Danida has chosen to support the project in spite of a number of very real points of criticism of the way the project has been run in previous phases:

- The project suffers from a marked lack of knowledge about the social and cultural situation in the area, and a corresponding lack of professionalism in its work with the local populations. The project has its strengths on the biological side.

- There has been a lack of real popular involvement in project activities and management decisions for the area.

- The project suffers from a lack of unity and coherence in its programme and activities. There is a need of better planning and coordination of the work.

The project is WWF-Denmark's first Danida-financed activity, and it holds a lot of challenges for us, particularly in 3 areas which are crucial for the succes of the project:

- We need to be able to interact constructively with the local people.

- We need to establish a technically reliable and yet simple and cheap system for monitoring of natural resources and biological diversity in a very large and inaccessible area.

- We need to establish a planning process which combines the scientific understanding with perspectives, understanding and needs of the local people.

In order to try to do this, we will be seeking cooperation with Danish experts and institutions and they are hereby invited.

Monitoring of biological diversity in Sahel
by means of satellite image analysis and ornithological surveys

Henning Nøhr & Steffen Brøgger-Jensen

The aim of the project is to develop a method for mapping and monitoring of biological diversity in the Sahel zone. Preservation of bio-diversity is an integrated component of sustainable exploitation of natural resources. Sahel is undergoing marked environmental changes, affecting not only human populations but also ecological communities. These ongoing processes have biological implications on both national and international levels. The Sahelian biotopes hold a number of endemic biological assets and constitute the main wintering ground for a significant number of Palearctic migratory bird species.

The project should ideally develop a tool which can be used for permitting implementation of planned strategies for the conservation and monitoring of bio-diversity.

The relations between the distribution of bird species and communities and the habitat types and land-use patterns will be elucidated by means of satellite.data analysis (CHIPS software, developed by the Inst. of Geography, University of Copenhagen) and standardized ornithological survey methods.

On the basis of the habitat selection patterns revealed in the analysis, a framework will be established for the use of bird species and bird communities as biological indicators to set priorities for conservation of bio-diversity in large geographical areas.

While carrying out ornithological surveys, this approach should at the same time ensure a further understanding of patterns and processes in the Sahelian ecosystems.

Census areas will be placed in Senegal and Burkina Faso, where ground surveys for satellite image analyses and botanical surveys have been conducted. The project thus aims to collaborate with ongoing, related activities, where these activities can contribute to descriptive and analytical information on the environment.

Male power and female participation in development programs
- The case of two West-African societies

Alain Lefebvre

The objective of this project is to contribute theoretically and empirically to the understanding of the power relations within the cultural construction of gender identity. It will examine the men's perception of the structure and functioning of gender ideology in two patriarchal societies (the Hausa and Mossi societies in *Niger and Burkina Faso* respectively), and the significance of this gendered subjectivity for women's participation in development programs.

The working hypothesis

In its effort to correct the male bias in the analysis of gender relations, feminist anthropology has shown that also in sexually inegalitarian social systems women possess certain degrees of economic and political autonomy, of authority and power. However, the analysis of the man's identity and its importance for the women's structural position in this type of societies has not received much attention in the feminist anthropology theoretical contribution. This is what this project intends to do. It attempts to reach a better knowledge of the power relations between genders by looking at them from the men's side. I argue that even though one has evidences of women's maneuverability in social, economic and sexual matters, their power is restricted to the specific female domain.

There are two reasons to this limitation. The first one is that the women's view is not an alternative to the patriarchal model but merely a tentative to become more visible within the cultural structures which assure their subordination. The second reason is that the patriarchal system is not a static and monolithic entity but an ideology with loose boundaries. It does not have an intrinsic essence but is the product of the male actors' constant manipulation and renegotiation of the reality so as to preserve their dominant position in the society.

The problematic of this project concerns in general the importance of ideology for promoting or hindering structural changes at the socio-political level, and in particular the knowledge of mechanisms which prevent women's power to challenge the patriarchal ideological values in their foundations. According to this assumption the women's participation in development programs is limited in patriarchal societies by the men's accept. These programs face the men's opposition if they bring a radical change to the traditional distribution of power between genders.

It has been observed that in both the Hausa and Mossi societies, men restrict women's participation in the struggle against desertification to the use of their labour force. They keep women excluded from land distribution, from agricultural informations, from the access to agricultural credit and machines, from decision-making in production cooperatives, from literacy programs and technical courses, from birth control policy, and the like. These difficuties have led to a whole set of government proposals to increase the women's involvement. They are all based upon a change in the rules of land ownership and upon a greater female presence in the decision-making process. In other words, measures which represent an intrusion into the men's domain. This tentative to improve the women's legal status have not succeeded to counterbalance the traditional conceptions of gender relations.

In my opinion, the reason behind the men's refusal to lose control over the realms of activities, which they considered to be their monopoly, is related to their fear of being deprived from an essential element of their identity, namely their power position. Men are opposed to a transformation of the ideas about the relations between genders and of their practices because they think that it could lead to their cultural alienation.

I assume that a legislation aiming at changing some fundamental aspects of the social order will fail if one does not take into account firstly the ideology which defines the traditional position and perception of each gender in the society and secondly the relations of this ideology with the rationality of the cultural system under consideration.

Aims of the project

In this work, gender identity and gender relations are considered as cultural constructs. They are elements of the people's meaning system which possesses its own logic.

The **first aim** of the project is to identify the perception(s) of the roles which the men have about themselves, about women, and about the gender relations. This will be achieved through investigating the forms of expression of the male cognitive universe which indicate what is the men's set f-definition, and their representation of women's identity and role in the society.

The **second aim** of the project is to analyze the legitimacy of this perception.

The majority of the population in both countries is Muslim. Consequently, reference to Islamic pres-criptions rule the people's behaviour in general and the gender relations in particular. I shall argue that they are based on a specific notion of personhood and of moral responsibilities expressed among other domains through a particular sexual ethics.

But the criticisms raised against the Orientalist school have shown that the functioning of Muslim societies is not only determined by religious principles. Islam does not provide a static, universal and a historical frame for all the actions and thoughts of the everyday life, especially in West Africa where its popular practice is still influenced by pre-Islamic animist values.

The **third aim** project is then to enquire how the day-to-day gender relations are also determined by other motives according to a cultural logic than the respect of the sacred texts. The perception of the gender ideology is part of the villagers's meaning system and as such is a cultural construct issued from specific historical, ideological and economic conditions. The project will therefore look at the relations between the schemes of interpretation and the very contexts in which individuals operate during three historical periods, the pre-colonial, the colonial and the post-colonial ones. It might show that the patriarchal system is susceptible to historical transformations and therefore open to a re-negotiation of the relations between genders.

The **last aim** of the project is to see the potentials for change in the traditional meaning system. The villagers are not individuals powerlessly submitted to ideological structures but social actors elaborating adequate strategies in their everyday experience. The study will investigate whether there exists various interpretations and manipulations of the given frame of gender symbols and sexual stereotypes. These alternative construction of the reality could be explained by the existence of social differences. Feminist anthropology has showed the importance of not considering women as an homogenous group. Disparities according to the age, class, status and kinship affiliation play an important role in the ways to perceive the reality. This conclusion is also valid for the study of men.

There might exist various male discourses among the Haussa and Mossi socio-cultural systems which are linked to the class relations. Some social classes and some individuals within them might have the power to monopolize the dominant values of the society and to formulate a discourse which became the dominant discourse. The analysis must therefore explain the relation between power and symbolical manipulation. It might be possible to locate loopholes in the ideological legitimation which could be exploited for an adequate adaptation of norms and customs.

The theoretical tools of the research

Theoretically, the project is confronted with the problem to find the relevant analytical tools which can fulfil a double task:

- firstly, the understanding of how a particular form of cultural representation -here the individual's identity- is experienced:

- secondly, the understanding of how this perception is produced and how it changes.

The concept of identity

The concept of personhood has been a recurrent subject within the Western philosophical thought, but in this study it is the work done by the phenomenological and hermeneutic approaches which is a source of inspiration. They operate within the same paradig-matic framework and complement each other.

Hermeneutics is concerned with the study of understanding through looking at the individual's interpretation of meanings and
intentions. In hermeneutics, the relationships between elements of human phenomena are perceived as mutually determining instead of being based on linear relations, and these relationships are defined in reference to the particular context wherein they are embedded.

Phenomenology is also concerned with meanings but more partic-ularly with their origin in people's thought. His founding father, E.Husserl, wrote that there do not exist any given objective facts but only perceptions of the world constructed in the individual's consciousness. Two forms of appropriation of a cultural meaning has been distinguished. One is called intentionality and refers to the knowledge resident within a subject; the other is called empathy and hints at the intuitive access to the thinking of another individual.

The multiplicity of the discourse

A cultural identity expresses norms and values which are collectively accepted. But this general adhesion is not similar to the absence of conflicts. The work of J. Habermas shows that the communication within a society does not always express the power relations which are hidden to the majority of social actors. Therefore the analysis of interpretation must be supplemented by a critical theory which unravels the non-perceived relations of dominance and the non-recognized deformations of the discourse.

It is known that one male discourse becomes dominant among all the others and acquires the ability to provide the people with explanations and definitions of themselves and others. This brings us to the necessity to understand the relationship between discourse and power. In this regard M.Foucault's argument that all discourse is a discourse of power will be taken into account for analyzing the strength of the dominant male discourse and its ability to tackle the contradictions which arise between the different male-male and male-female perceptions.

Methodology

The project has a 3 years duration with the following provisory work-schedule:

- 2 months pilot project with a visit to Burkina Faso and Niger. I have applied for a travelling grant at Uppsala Afrika Institute.

- 6 months for the reading of theoretical literature about identity. I would like to join the Program in the History of Consciousness at the University of California, Santa Cruz for one semester.

- 6 months for the reading of literature about the Haussa and Mossi societies.

- 10 months fieldwork in Burkina Faso and Niger.

- 12 months for the writing of the report.

Locust control in Africa
- The New Role of Malathion

L.E.K. Pedersen and Kristian Lystbaek

Grasshoppers especially the Senegalese grasshopper, Oedaleus senegalensis (Krauss, 1877) - is considered an important pest making life difficult for farmers in large areas of West Sahel. Assistance to grasshopper control could therefore be a very effective means for support of local agriculture and food production.

Our activities are summarized in the following.

Introduction

Locust plagues have been a recurrent menace to human welfare in Africa through the history of civilisation (Gutsch, 1987). Invasions by swarms of the Desert Locust *Schistocerca gregaria* (Forskål, 1775) are part of the ecology not only in the Sahara zone but also in much wider areas. The invasion area covers more than 20% of the total land surface of the globe (Steedman. 1988). A recent transatlantic flight to the Caribbean Islands illustrates the extraordinary migration power of the Desert Locust (SAS.1988). Evidently the potential for crop destruction by the Desert Locust is very substantial. Although hard data are rare, estimates indicate that locust plagues constitute a very serious threat to agriculture in the affected areas (Skaf, 1986: Musuna. 1988: Steedman. 1988).

Chemical control

In areas plagued by locusts and grasshoppers chemical control is used frequently to control the pest population. A variety of insecticides have been used with different preferences in the different regions (McEwen, 1982: Symmons, 1984; USDA, 1987: Brader, 1988). Over the last decade or longer the general trend has been a shift away from persistent organochlorine compounds to less toxic and less persistent insecticides. Several products are generally acceptable for use in locust and grasshopper control (Launois-Luong et al. 1988) and by irregular intervals FAO issues a list of products in use in well organized control operations. The insecticide Dieldrin has played a central part in locust control over many years. This insecticide has proven extremely efficient in the barrier spraying techniques developed in the 1950,s for the control of hopper bands.

High toxicity to mammals, long persistence in the environment, and a large potential for bioaccumulation have led to an almost general ban on Dieldrin. making it unacceptable to the major donor agencies during the 1988 campaign. The fact that a limited initial outbreak had quickly developed into a more or less general pan-African plague in 1988 has led to speculations whether usage of Dieldrin could have stopped the invasion in due time (Skaf. 1988a. 1988b). Other important elements played their part in the initial control failure: deterioration of the local locust control organizations, inaccessibility of key areas and civil war hampered the control effort significantly (Brader. 1988).

During the 1988 control campaign which was heavily supported by the international community, emphasis has been placed on chemicals 'that degrade readily in the environment (Brader. 1988). Among the insecticides presently used. Malathion has gained a dominant role in African Desert Locust control (SAS, 1988; Pedersen, 1989). This insecticide is relatively new in large scale locust control in Africa. It is among the very few insecticides registered for and extensively used for control of grasshoppers in the USA where the product is well-known for its efficacy (USDA, 1987). Malathion is now considered a standard insecticide for use against grasshoppers as well as locusts in Africa. (SAS, 1988; Launois-Luong et al., 1988). Control practice in Africa has proved that it is not only very successful against the Senegalese grasshopper (Walsh, 1986a. 1986b), but also against the Desert Locust. In the Maghreb countries in northwestem Africa the locust control campaign has been outstandingly well organised (Lorelle, 1989). During the massive spring campaign of 1988 against the Desert Locust in Morocco, Algeria and Tunisia, Malathion has become the preferred insecticide (Pedersen, 1988). In this area the spray operation has been particularly successful: in spite if a very heavy locust infestation little crop damage has been reported. In Mali and Senegal Malathion spraying has also played a central part and in these countries the control effort has been considered successful (CILSS, 1989). According to the authorities in the affected countries, the main reasons for the increasingly important role of Malathion among several effective acridicides are:

1. The product has a very low mammalian toxicity. This is very important in terms of protecting the pest control operators in large scale campaigns where safety precautions may not always be up to international standards.

2. The product is inexpensive (despite its being a highgrade technical chemical of more than 96% purity) and is one of the most widely used insecticides in agriculture and public health and therefore readily available from reliable sources.

Additionally it is well known that Malathion ULV formulations are easy to handle and that the product is a non-persistant, non-bioaccumulating chemical. During the control campaign in 1988, 14-15 million ha. were sprayed with insecticides (SAS 1989). This unprecedentedly large effort has raised the following key questions:

1. How do we prevent locust plagues? How can locusts be controlled at an early stage, before the situation develops into a disaster?

2. What is the impact of the anti-locust chemical control measures on human health and the environment?

The answer to the first question obviously is related to improved surveillance, early warning, and international collaboration, all based on strong and efficient local control organizations (Brader 1988, Steedman 1988). This is uncontroversial in theory but very hard to practice continously) in Africa. A further discussion along these lines is beyond the scope of the present paper. Concerning the second question, as a result of the overwhelming toxicological and environmental concern in the Western community, pesticides have been subject to a very thorough test scheme. This makes pesticides one of the best scrutinized group of chemicals - often more information is generated for pesticides than for cosmetics and food additives (Cedar, 1987). Since Malathion has been in use in crop protection and public health for more than three decades, a substantial amount of experience on the safety of this product has accumulated.

To answer the question of the effects on human health and the environment, the properties of Malathion are reviewed below in relation to locust and grasshopper control in general.

State and trends in the flora and vegetation
of the Noflaye Reserve (close to Dakar, Senegal)

Jean Baptiste Ilboudo

The Noflaye Reserve is located about 40 km Northeast of Dakar. It was created in 1957, and its present area is nearly 16 ha. Adam (1957) made floristic inventory in this forest, and ennumeratcd 376 herbaceous and woody species.

Unfortunately the degradation of the climatic conditions in Senegal and the strong human impact have affected the reserve severely. Our study is aiming to re-establish and conserve this botanical reserve and so make it more valuable for the social and economic development. The study includes the following components:

1. Knowledge about the present flora and vegetation.

2. Find the present trends in the vegetation and compare with earlier evidence.

3. Find the principal factors for the dynamics of the vegetation, its potential use for the surrounding human population.

4. Propose a type of management scheme, making the reserve a scientific site to study local plant species and to save rare and endangered species, and also to make Noflaye a part of various types of education.

Comparative studies of the woody flora and vegetation of Senegal
- summary of activities.

Jonas E. Lawesson

The Aarhus University Sahel project at Botanical Institute was initiated in 1988, aiming at study the composition, structure and distribution of woody communities in Senegal (JEL). Later pollination and seed predation studies of Acacia species were added (Knud Tybirk). Studies of Combretaceae in Burkina Faso have been carried out (Jens Koed). Detailed phytosociological studies are being carried out in the Nofflaye forest reserve, close to Dakar (Jean Baptiste Ilboudou). Further activities on the ecology of bush fires, and detailed studies of National parks are planned as well.

So far, flora and vegetation structure, population characteristics and phytogeography have been studied at 63 classified forest or reserve locations covering the Sahelian, Sudanean and Northern Guinean ecoclimatic zones (see map below, fig.1.), excluding Basse Casamance. The savanna zone was investigated with strip transects, circular hectare plots and Nearest Neighbour methods. The forest zone was studied with quadrangular 10 x 10m quadrats. General species lists were also made at several locations. In depth floristic and vegetation studies were carried out in 3 little known areas: the national parks of the Saloum Delta, Niokolo Koba and the mountain areas of SE Senegal.

The data collected were subjected to numerical treatment mainly Detrended and Constrainted Correspondence Analysis, with precipitation and pedology as the main constrained axes. Based on ordination and the use of classification algoritms in a Two Way Indicator Species Analysis, the available data were classified in a number of vegetation types, which are related to phytogeographical elements of West Africa. A comparison with classified NOAA images is attempted. Distribution maps of most the woody species studied in Senegal (close to 200) are being prepared, based on own data and literature sources, and their phytogeographical affinities given. The population characteristics of the same species are studied, and used to deduce possible trends in the vegetation.

Figure 1: Study sites 1988-90

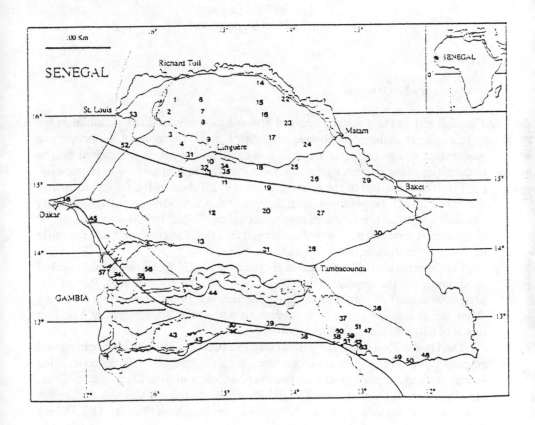

Summary of Danish Red Cross Activities
in the Sahel Region

Danish Red Cross
International Department

1. Sudan

a. Veterinary Programme

The civil war in the sourthern part of Sudan has eliminated the possibilities for large segments of the population to earn their living. Traditionally, the people are pastoralists whose cattle, in the arid and semi-arid areas, is dependent on the possibilities of moving over great distances in order to find grazing opportunities.

The fragmentation of the country in areas controlled by the Government and the rebel forces respectively and the resulting deterioration of the security situation has made this freedom of movement impossible. In addition, virtually all Government services, including the veterinary service, has collapsed, and the cattle has become increasingly vulnerable to disease.

The International Committee of the Red Cross (ICRC) runs a multifaceted programme for the benefit of the victims of war in the conflict area. Part of this programme aims at establishing an emergency veterinary service which among other things implies vaccination against rinderpest of approximately 4 million heads of cattle in the entire area.

The Danish Red Cross has agreed with the ICRC to take over the operational responsibility of the veterinary programme, meaning that all decisions concerning funds, staffing, reporting etc. is the responsibility of the Danish Red Cross Headquarters in Copenhagen. The field work is carried out through the Danish Red Cross delegates based in Lokichokio in Northern Kenya. The Danish delegates who move frequently into the conflict area are, however, required to observe all security regulations which remain the responsibility of the ICRC.

The project which is financed by Danida has a budget of DKK 28 million and was initially planned for 15 months dating from 1 October 1989. However, delays in the activities occurred in periods during 1990 because of increased hostilities in the area, making operations impossible. Consequently, the programme has been prolonged until the end of 1991.

b. Environment and Water Development Programme in the Derudeb Area

About 300 km from Port Sudan on the Western side of the Red Sea Hills, the Danish Red Cross has agreed with the Sudanese Red Crescent Society to implement a project, which should contribute to alleviate the effects of several years of drought for the local population of some 35.000 people, and to reduce the dependence on relief. The population in the area is predominantly nomadic, and the traditional life style has been threatened by the drought to a point where migration to the towns could easily be foreseen.

The aim of the project is to contribute to an improved water supply by restoring existing wells, which have decayed, at the same time examining the geological possibilities for establishing new wells and other water sources to be constructed in a subsequent phase of the project. One Danish delegate has been attached as a Project Coordinator for the local branch of the Sudanese Red Crescent Society, which has the direct responsibility for the implementation.

The project has been planned initially for two years as from 1 April 1990 with a subsequent phase of five years. The activities have been budgeted to DKK 9 million. Danida has agreed to finance the entire project.

2. Senegal

In the Senegal River Valley, at the locations Maka-Diama and Podor situated appr. 50 km and 200 km respectively up river from St. Louis, the Danish Red Cross has since the beginning of 1987 supported two small-scale vegetable garden projects. The projects are being implemented through the local committees of the Senegalese Red Cross with logistical support and professional advice from the headquarters in Dakar.

The overall aim of the project is to contribute to attaining self-sufficiency in food, to contribute to restoring the vegetation cover and to counteract the tendency to migration towards the cities. The direct target group is a total of 70 families, selected by the local committees of the Red Cross along certain needs criteria. Each family is given a plot of land of appro. 1,000 square metres put at the disposal by the local municipality, and the families are producing vegetables according to an agreed plan. The produce is partly being consumed, partly sold on the local market, and a proportion is being paid into a revolving fund with the aim of attaining economic sustainability over a period of years.

The present agreement between the Senegalese Red Cross and the Danish Red Cross runs until mid-1993, and the projects, which have a budget of appro. DKK

1 million over three years, are being financed by the Danish Red Cross branches in Southern Jutland.

3. Regional Environmental Education Project in Burkina Faso, Senegal and Sudan

On an idea originally conceived by ENDA-Tiers Monde, Dakar, the Danish Red Cross has embarked on an environmental education programme, which has initially been started in the three above-mentioned countries in the Sahel area.

The long term aim of the project is to contribute to a greater awareness among the population of the Sahelian Belt of the factors threatening the environmental equilibrium, with a view to making the population more capable of managing the natural resources in an economically and ecologically viable way.

The more immediate aim of the project is to introduce in the schools and among the children and youth outside the school system an environmental education curriculum, which is adapted to the daily life of the children. The education should draw on both traditional knowledge and modern scientific methods, and the project should follow a participatory approach.

In each of the countries of intervention, a number of schools and villages in areas threatened by desertification have been selected, and teachers and specially selected "village animators" have been trained. Teaching material is being developed, and in order to illustrate the theoretical training on environmental issues, certain specific small scale projects are being undertaken. Such projects are typically tree-planting, intensive breeding of sheep, health-related activities etc.

The project is being implemented through the national Red Cross/Red Crescent Societies in the respective countries. The societies Collaborate in the execution of the projects with the local ministries concerned as well as with NGO's, such as END, Dakar, and Six-S, Burkina Faso.

The project was originally planned for appro. 20 months in Burkina Faso and Senegal dating from 1. January 1989. In Burkina Faso, a second phase is currently under preparation, and in Senegal the question of an extension is being considered. In Sudan the project is still in its initial phase running from 1. January 1990 to 31. December 1991.

The project has so far been financed by the proceeds from the 1986 sales of the "Childrens Third World Calendar", and had in the initial phase for all three countries combined a budget of DKK 5 million.

4. Emergency Relief

In addition to the above mentioned projects which are all carried out in a bilateral relationship between the Danish Red Cross and the national Red Cross/Red Crescent Societies, the Danish Red Cross is continuously supporting the emergency relief activities of the ICRC and the League of Red Cross and Red Crescent Societies.

The contribution from the Danish Red Cross may be in both cash, kind (purchase and shipment of relief goods, where local purchase is not possible) and services (recruitment of Danish delegates who are being put at the disposal of the League or the ICRC).

Monitoring and evaluation of local people's participation in Renewable natural resource management projects in the Sahel : A case study of the Sand Dune Stabilization and Agro-Sylvo-Pastoral Development Project in Mauritania.

Søren Lund

Summary

Assessments made of renewable natural resource management projects in the Sahelian region towards the mid-eighties generally showed results to have been largely insufficient. The modest levels of local people's participation in project activities was evoked as being one of the key critical points of concern.

One of the major projects of this kind is the Sand Dune Fixation Project in Mauritania. Implemented by the FAO, the project was launched in 1983 on a joint UNSO/DANIDA, UNDP, WFP, and UNCDF funding. In the course of its first phase, 1983-1986, the Sand Dune Fixation Project achieved positive technical results and developed a contractual model of collaboration with the local communities.

The project has been continued under the name of the Sand Dune Stabilization and Agro-Sylvo-Pastoral Development Project during a second phase running out in 1991, based on this new contractual model of collaboration and testing out elements of an integrated agro-sylvo-pastoral approach. It is now possibly being extended into a third phase.

The main questions underlying this research project from the outset has been what makes local people participate in such activities. Based on the experiences gained during the first phase, the research design was originally intended as a contribution towards the elaboration of an M & E model concerning the local people's participation in project activities during the second phase of the Dune Stabilization Project in Maritania. The purpose was to explore what model of "participation" should be applied during project implementation on order to assure the sustainability of project activities (sand dune fixation) after the expiration of the external project funding and technical assistance.

As the research project has advanced, attention has been switched into the direction of understanding the local people's participation as an expression of the relationship between local actors and public administration as it manifests itself through local economical, political, and cultural institutions, and as a reflection of the prevailing social structures.

The last field work to be conducted during March and April 1991, concerns the importance of local management regimes, taken as systems of rules and work-practices, in providing incentives for the individual landmanager to "participate", i.e. to invest labour and capital in sustainable management models.

3. Summaries and discussion notes
(the experiences and strategies of development agencies)

Danida

Presentation by Hans Genefke
Summarized by I. K. Higashidani & S. Leth-Nissen.

Danida has focussed on Niger and Burkina Faso as potential countries for cooperation in the Sahel Area. The Danish Government has not yet (1991) taken any decision on which countries are to be the main cooperative countries. This summary will shortly focus on Danidas experiences and strategies with the above mentioned countries.

Danida's overall objective in Sahel:

To rehabilitate, preserve and develop the natural resources in the region with the view to increase the production, especially the production of foodstuffs.

Secondary objectives:

* To support the possibilities within the region to finance development plans through increased exports.
* To release the stress of the natural resources through the support of the energy sector.
* To improve the living conditions for the poorest (nutrition, water, health, education).
* To improve the conditions for women.
* To reduce the growth of the population.
* To contribute to development of democratization and local public participation.

The last point is still getting a more and more important issue for Danida. The new approach is the integrated multi-sectoral projects with environmental elements for rural people. Also a stress is given the institutional strengthening as well as the social infrastructure and human development in general. According to the Human Development Index (cf. UNDP's Human Development Report 1990, 1991, 1992) the sahelian countries are among the weakest of all.

Danida priorities in Sahel:

It is most likely that Danida will choose approximately 20 countries for bilateral cooperation. At the end of 1990 the first 12 was chosen. The rest will probably be chosen within the next 5 years. In the meantime Danida is preparing analyses and strategies for the potential countries. This has also been made for Niger and Burkina Faso.

On the Report concerning "Perspectives for Danish development cooperation with Niger:

Contents:

1. Main characteristics of the economic, social and political situation of Niger.
2. External relations.
3. Diversity and dynamics in the society of Nigeria.
4. Main issues in a national development.
5. The strategy of the Niger Government.
6. External assistance
7. Perspectives for Danish Development cooperation with Niger.

Approach: Basic principles for the Danish cooperation.

* Geographical concentration.
* Multi-sector projects.
* Institutional strengthening.

Possible future interventions:

* Multi-sector projects
* Institutional strengthening.
* Research activities
* Financial support to other integrated projects.
* Infra-structural activities at national level.

Basic devevelopment problems:

* Severe and increasing economic problems.
* Environmental problems and deterioration of natural resources
* A fragile food security situation
* Underdevelopment of human resources (HDI)

Overall objective for Danish development cooperation:

To rehabilitate, preserve and develop the natural resources in the region with the view to increase the production, especially the production of foodstuffs.

Secondary objectives:

* To improve the living condition for the weakest and most vulnerable parts of the population (nutrition, water, health, education and food security).
* To release the pressure on the natural resources and promote a rational use of the limited natural and economic resources.
* To support the dev. of the national research capacity, particularly as concern the sustainable utilization of the natural resources.
* To reduce the growth of the population
* To support the dev. of democrati and local public participation.
* To release the ethnical tensions within Niger.
* To contribute to an increased trade among the countries in region.

Future bilateral development cooperation with Niger - a preliminary draft.

a. A multi-sectoral approach in Zinder and Diffa.

* A continuation of existing project for drinking water introducing new sectors in the project.
* Interim support for the former WB livestock project with a view to a continuation into a natural resource management project.
* Realisation of a project for food security adapted utilization of the natural resources

* Technical assistance/institutional strengthening to physical planning (national, depart., arrondissement)
* Rehabilitation of feeder roads
* Support national activities related to these activities:

    - cellule du PIGRIN
    - INRAN
    - Early warning systems

b. A sector-oriented cooperation an national level.

* Health care programme
* Technical assistance/institutional strengthening (national, departmental).

On the report concerning "Perspectives for Danish development cooperation with Burkina Faso."

Contents:

1. The regional and social development
2. The political situation
3. The economic development
4. The development of the sectors
5. The development strategy of the government
6. External assistance to Burkina Faso
7. Perspectives for Danish Development cooperation with Burkina Faso

Objectives:

* Development of sustainable rural production systems
* Improvement of social and physical infrastructure
* Strengthen the aims towards democrati and decentralization

Approach: Basic principles for the Danish cooperation.

*   Geographical and sectorial concentration
*   Strengthening the weakest groups of the population
*   Institutional strengthening

Possible future interventions:

*   Multi-sector rural projects in the eastern region (ecologically sustainable)
*   Physical infrastructure (rural water supply, feeder-roads as support to multi-sector approach in selected geographical areas)
*   Social infrastructure (health-sector, education, supply of medicine)
*   Urban infrastructure (power generators and water supply in secondary centers)
*   Democracy (local radio/broadcasting, education of journalists, Human Rights-institutions)
*   Twinning with Burkinnabe research institutions

The European Community

Presentation by J. L. Baudet
Summarized by I. K. Higashidani & S. Leth-Nissen.

Micro-economics related to EC and Sahel was mainly presented. First some general points on development problems. It was stated that development is a long term process and the necessity of a dialogue between governments and organizations was also stressed.

As 90 % of the Sahel-population live in rural regions, EC tries to strengthen two aspects: 1) food security and 2) development of the agricultural sector. Development in the agricultural sector has been concentrated on development of agricultural productivity, but to improve the living conditions in the rural areas, EC must take a broader approach in contributing to development of market relations between rural and urban areas.

Three components to discuss in order to try to achieve this objective:

1) Development of internal conditions.

The mining is bad, the natural resources are over-exploited and the extensive cultivation of new land is going on instead of intensification of existing agricultural land. Import and food aid also contribute to low prices and low productivity.

What are to be done to ensure sustainable agricultural growth?
a) resource base on landrights.
b) irrigation programmes.
c) use of fertilizers.
d) land tenure systems are to be changed in order not to slow down agricultural development.
e) better prices.

2) International relationships.

We have to create methods to deal with the problems concerning international relationships.

a) cereal production is to be strengthened on the national scale, but on the international scale, this cereal production is obstructed by very low prices from the world market. This means that the incentive to produce more cereals to feed the national market is not present.

b) livestock. EC price policy means that meat prices are very low in Africa and this obstructs any internal development in Sahel.

3) Non-agricultural issues.
The non-agricultural issues are mainly related to the demographic situation. EC want to support at grass-root-level, local organization by the Africans themselves.

a) local ability to buy
b) structural adjustments

Discussion:

? Comment: Democratic structures are not a question of either/or. You can do both. Dialogue is important.
! EC will support democracy and assist in development of grass-root organizations.

? Import of cereals is mainly rice and wheat. It is by now difficult to turn back to sorghum and millets. The tyranny of rice in Sahel is going on, a former director of CILLS, has stated.
! I agree with that. Maize has the best potential but perhaps there are limitations due to negative preference from people.

? On prices, please comment.
! Regulations are needed.

? No further acceptance of dumping cereals or food stuffs from EC.
! It is necessary to improve livestock production even though markets are flooded in Sahel/Africa. Therefore considerations on future reforms in EC concerning these things.

? On tyranny of rice: one half of the budget from EC went to irrigation for rice in Mali and Niger. How does this correspond to the need, the Sahelians have?
! There is a potential for these crops in these countries.

The IUCN approach in the sahel

Presentation by Per Rydén
Summarized by Eva Bierrum Madsen

The IUCN Sahel Programme is operational since 1988. It covers 10 countries from Mauritania in the west to Somalia in the east. The first three years of Programme operation have been a learning process and characterized by finding the place for IUCN in the community of organizations which have worked in the region for much longer. It has been a question of merging the mandate of IUCN - "to work for achieving harmony between humanity and nature" - with the needs and aspirations of Governments and peoples of the region, also taking into account the experience gained by other organizations.

In a workshop, at the 18th General Assembly of the IUCN, experiences from different arid regions of the World were exchanged and discussed. The Sahelian region exhibits some very important differences from the other arid regions discussed (Botswana, Saudi Arabia and Australia) especially with respect to the poverty of the region and the strong dependency on the natural resource base; the entire socio-economic structure is also less developed in the Sahel than in the other regions. Similiarities between the Sahel and the other arid regions refer to the functioning of the ecosystem, in particular to the highly unreliable rainfall. It was pointed out that strategies in such areas should be event-oriented rather than based on long term planning. The challenge is to find production systems that allow for a maximum exploitation of the natural resource base but which do not cross natural limits so as to result in degradation.

The role of IUCN as an international environmental organization is to contribute to a better understanding of resource management problems in the Sahel. Some IUCN studies have been carried out in the region and results and conclusions of these studies have been published in the first volume of "The IUCN Sahel Studies, 1989" Since many years IUCN has been working with National Conservation Strategies as its contribution to formulating resource management strategies at the national level. The NCS concept, as one example of Strategic planning, is a vital component of the IUCN Sahel Programme. All IUCN projects in Sahel are concerned with the management of protected areas.

Studies and field projects aim at contributing to the search for sustainable resource management systems in the Sahel. As IUCN is not a development or donor agency the role of the organization is mainly catalytic, working in partnership with local institutions, governmental as well as NGOs. IUCN seeks to reinforce local NGOs by providing information on conservation or ecological principles, either by having local organizations as partners in field projects or by providing information and training through workshops or more permanent support to NGO umbrella organizations.

It is recognized by the IUCN that there is a need for a new approach to training.

Some blending of local and external knowledge is needed when addressing natural resource management problems. Instead of attempting to "sell" ready solutions to the farmers the Sahel Programme tries to pursue a "loop model". The tasks of this model are firstly to understand the logic and ecological stability of traditional resource management systems, secondly to investigate why traditional practices are no longer adequate and to identify areas where management has to be adjusted, and thirdly to develop potential innovations which can be tested in the field. This loop process requires intensive communication between the local population and the external agents.

IUCN aim at establishing an advisory group for the Sahel Programme. The local point of such a group should be in the Sahel itself, but external expertise can and should play a role. The IUCN Sahel Programme therefore seek partners among specialists with considerable experience in the Sahel.

CARE - philosophy and projects in sahel

Presentation by Remko Vonk
Summarized by Eva Bjerrum Madsen

CARE is an international NGO focussing its activities on projects concerning agriculture and natural resources, small economic activity development, and primary health care. CARE has been working in Sahel since 1974. At present, activities are carried out in Niger, Chad, Sudan, Mali, Egypt, Somalia, Cameroon, Ethiopia and Togo. Most of the present activities evolved from food aid programmes after the 1984 drought. The activities are carried out at community level and primarily for the benefit of poor people.

The activities of the 'agriculture and natural resources' division have included forestry with farmers, sand dune stabilization, agroforestry, natural regeneration and agriculture. The objective of the projects has evolved towards creating an increased land-based economic security.

The approach is one of participation, sustainability and A/G TV (Amenagement/Gestion de Territoire Villageois). For participation it is important to differentiate between projects which are within direct interest of the community or the individual and projects which are beyond direct personal interest. For example, participation is not always easily obtainable in terms of tree planting because there is no immediate interest or 'reward' for local participants. This in turn may result in a lack of sustainability of the project. Such projects may have to be carried out as 'public works' where the participants are paid in money or food.

Sustainability may be obtained at different levels: technology, farmers' know-how, farming system, community and community support organizations. It is not the project that should be sustainable but the result.

The A/G TV is a governmental planning method driven by CILLS/Club du Sahel and widely adopted by West African governments. CARE projects are a part of A/G TV. The model is not always useful for NGOs because it is characterised by taking the village as one homogeneous unit whereas NGOs may work with interest groups other than the village (e.g. young people, women, pastoralists).

NG0 projects tend to evolve from technology driven to integrated area defined programmes gradually increasing the number of different activities within the area. This evolvement often results in a conflict of interests between the number of activities within the project and the often limited capacity of such organizations.

Notes from concluding sessions

Lessons learnt

The concluding sessions of the Workshop centered around three major topics: (i) desertification, (ii) utilization of natural resources, and (iii) participatory models.

The concept of desertification needs to be clarified and made more operational. Specific indicators of desertification exist, e.g. rainfall, soil erosion, dust storms, soil fertility in relation to current agricultural yield and natural biomass production, and changes in the extent and composition of the vegetation cover. For most of the indicators there is, however, a lack of factual knowledge of both short and long term trends of development. It was discussed whether social factors should be considered indicators of desertification.

The understanding of institutional, legal and tenure aspects, and user rights in natural resource utilization is often lacking. Local knowledge, no doubt, has an important role to play in providing or improving such an understanding. Appropriate technology has not been sufficiently well developed and maintenance is often a major problem. Credit schemes to support local communities to maintain locally developed technology is one possible way ahead, whereas environmental rehabilitation often is likely to be beyond the scope of such credit schemes.

Decentralization seems to be the keyword for a succesful implementation of participatory models. It was, however, recognized that too often the state becomes the "scapegoat" for failed projects. Better co-operation with local NGO's, local communities, private local firms, and individuals is needed, nevertheless the state should be taken seriously. There is not always opposition between the state and the people, although it does occur.

Workshop Programme

MONDAY, JANUARY 7TH, 1991

Theme 1 : The context
- Review of Sahelian problems and solutions. Sofus Christiansen, Institute of Geography, Copenhagen.

Theme 2 : Experience and strategy of funding and executing agencies.
- Danida Hans Genefke.
- UNSO   Peter Branner, Director/Henrik Secher Marcussen, PTA.
- EC   J.L.Baudet, Sahel Division, DG VIII.
- IUCN   Per Rydén, Sahel Programme Coordinator.
- CARE   Klaus Branner Jespersen, Danmark.

TUESDAY, JANUARY 8TH, 1991

Theme 3 : Degradation
- Desertification - status and problems. Ulf Helldén, Institute of P h y s i c a l Geography, University of Lund.
- Soil erosion in Lesotho. Lennart Strömqvist, SAREK, Uppsala University.
- Desertification control on sandy soils in Sahelian West Africa with special reference to Mauritania : 1. Ecological problems and technical solutions, 2. Socio-economic bottle necks. Axel Martin Jensen.

Theme 4 : Socio-economy
- Village associations and the state in Senegal. Henrik Nielsen. International Development Studies, Roskilde University Centre.
- Land tenure and agro-pastoralists in Senegal. Kristine Juul, Centre de Suivi Ecologique, Dakar.

Theme 5 : Water
- Water management. Jean-Pierre Zafiryadis, I/S Krüger.

Theme 6 : Forestry
- Natural forest management by local people in Burkina Faso. Per Christensen, FAO.
- Non-poisonous methods of pest control during seed collection, handling, storage and transfer. Henry Mourier, Danish Pest Infestation Laboratory.

- Forest policy in Senegal. Bienvenu Sambou, Institut de Science de l'Environnement, Université de Dakar.

Theme 7 : Exchange of experience
- Interdisciplinary group sessions

WEDNESDAY, JANUARY 9TH, 1991

- Lessons learnt and recommendations
- Evaluation

Workshop Participants

Alstrup, Vagn, Institut for Økologisk Botanik, Øster Farimagsgade 2D, 1353 København K
Andersen, Henrik Steen, Geografisk Institut, Øster Voldgade 10, 1350 København K
Baudet, J.L., CE Sahel Division, Rue de Loi 200, 1049 Bruxelles, Belgium
Bregengaard, Per, Wesselsgade 2, 1th, 2200 København N
Brimer, Leon, Institut for Farmakologi og Patobiologi, Bülowsvej 13, 1870 Frederiksberg C
Buch, Jette, Danida DB.5, Asiatisk Plads 2, 1448 København K
Brøgger-Jensen, Steffen, Ornis Consult, Vestergrogade 140, 1620 København V
Carstensen, Kim, Verdensnaturfonden, Ryesgade 3F, 2200 København N
Christensen, Per, Skovrider, Østjysk Skovdyrkerforening, Låsbyvej 18, 8660 Skanderborg
Christiansen, Sofus, Geografisk Institut, Øster Voldgade 10, 1350 København K
Damgaard-Larsen, Søren, Danish Red Cross, 28 Dag Hammarskjölds Allé, 2100 København 9
Degnbol, Tove, COWIconsult, Parallelvej 15, 2800 Lyngby
Falk, Jo, Pottemagertoften 131, 8270 Højbjerg
Frederiksen, Peter, RUC, Postbox 260, 4000 Roskilde
Friis, Ib, Botanisk Museum, Gothersgade 130, 1123 København K
Furu, Peter, Danish Bilharziasis Laboratory, Jægersborg Allé 1 D, 2920 Charlottenlund
Genefke, Hans, Danida, Asiatisk Plads 2, 1448 København K
Grant, Stewart, Hedeselskabet, P.O. Box 110, 8800 Viborg
Graudal, Lars, Danida Forest Seed Centre, Krogerupvej 3A, 3050 Humlebæk
Hansen, Mogens Buch, RUC, Inter. Dev. Stud., Postbox 260, 4000 Roskilde
Higashidani, Ingeborg K., Institut for Etnografi og Social Antropologi, Moesgård, 8270 Højbjerg
Hellden, Ulf, Natur Geografiska, Lund Universitet, Sølvgatan 13, 2263 Lund, Sverige
Heim, Claudia, Stud.IU, RUC, Thorshavnsgade 10, 2300 København S
Ilbodou, Jean Baptiste, Botanisk Institut, Nordlandsvej 68, 8240 Risskov
Jensen, Axel Martin, Consultant, Poutet en Bas, Daumazan sur Arize, 09350 France
Jensen, Dorrit S., I. Krüger, Gladsaxevej 363, 2860 Søborg
Jensen, Kurt Mørck, Danida DB.5, Asiatisk Plads 2, 1448 København K
Juhl, Kristine (CSE), C/O Ida Juhl, Vesterbrogade 21, 1620 København V
Keiding, Henrik, Danida Forest Seed Centre, Krogerupvej 3A, 3050 Humlebaek
Kieler, Jan, CARE, Borgergade 14, 1300 København K
Koed, Jens, Botanisk Institut, Nordlandsvej 68, 8240 Risskov
Kress, Marianne, Stud. IU, RU, Olof Palmesgade 6, 4th, 2100 København
Krogh, Lars, Geografisk Institut, Øster Voldgade 10, 1350 København K
Lawesson, Jonas Erik, Botanisk Institut, Nordlandsvej 68, 8240 Risskov
Lefebvre, Alain, Nykongensgade 9, 1472 København K
Leth-Nissen, Søren, Engkærgårds Allé 108, 8340 Malling
Lillelund, Hans, Danida, Asiatisk Plads 2, 1448 København K
Lund, Søren, RUC, Inter. Dev. Stud., Postbox 260, 4000 Roskilde
Madsen, Eva Bjerrum, Botanisk Institut, Nordlandsvej 68, 8240 Risskov
Marcussen, Henrik S., UNSO, One UN Plaza, New York, N.Y. 10017 U.S.A.
Meyer, Marlene, Geografisk Institut, Øster Voldgade 10, 1350 København K
Michelsen, Jette, c/o Tybirk, Botanisk Institut, Nordlandsvej 68, 8240 Risskov
Michelsen, Anders, Botanisk Museum, Gothersgade 130, 1123 København K
Mikkelsen, Troels, Danish Red Cross, 28 Dag Hammarshjölds Allé, 2100 København 9
Moestrup, Søren, Danida Forest Seed Centre, Krogerupvej 3A, 3050 Humlebæk

Mourier, Henry, Statens Skadedyrslaboratoriet, Skovbrynet 14, 2800 Lyngby
Müller, Jens, Cheminova A/S, Postbox 9, 7620 Lemvig
Nielsen, Flemming, Geografisk Institut, Øster Voldgade 10, 1350 København K
Nielsen, Henrik, c/o Ida Juhl, Vesterbrogade 21, 1620 København V
Nielsen, Ivan, Botanisk Institut, Nordlandsvej 68, 8240 Risskov
Nøhr, Henning, Ornis Consult, Vestergrogade 140, 1620 København V
Rüdinger, Erik, Damida DS.III, Asiatisk Plads 2, 1448 København K
Sambou, Bienvenu, Université de Dakar, Institut de Science de l'Environnement, Dakar, Senegal
Schou, Inge, Sølvgade 34, 4, 1307 København K.
Schønnemann, Jørgen, COWIconsult, Parallelvej 15, 2800 Lyngby
Smida, H., Commission des Communautes Europeennes, Rue de la Loi 200, 1049 Bruxelles, Belgium
Speirs, Mike, Danagro Adviser, 8 Granskoven, 2600 Glostrup
Strömqvist, Lennart, SAREK, Universitet Uppsala, St. Olofsgatan 1lb, 753 21 Uppsala, Sverige
Søndergaard, Povl, Arboretet, 2970 Hørsholm
Poulsen, Ebbe, RUC, Inter. Dev. Studies, Postbox 260, 4000 Roskilde
Rasmussen, Kjeld, Geografisk Institut, Øster Voldgade 10, 1350 København K
Reenberg, Anette, Geografisk Institut, Øster Voldgade 10, 1350 København K
Ryden, Per, IUCN, Avenue Mont Blanc, 1196 Gland, Switzerland
Vejlgård, Birthe, Fælledvej 15, 3, 2200 København N
Zafiryadis, J.-P., I. Krüger, Gladsaxevej 363, 2860 Søborg
Zeuthen, Kristine, Geografisk Institut, Øster Voldgade 10, 1350 København K
Ørum, Torkild, I. Krüger, Gladsaxevej 363, 2860 Søborg